Programming the Intel® Galileo

Getting Started with the Arduino™-Compatible Development Board

Christopher Rush

Mc
Graw
Hill
Education

New York   Chicago   San Francisco
Athens   London   Madrid
Mexico City   Milan   New Delhi
Singapore   Sydney   Toronto

Library of Congress Control Number: 2016955477

Programming the Intel® Galileo: Getting Started with the Arduino™-Compatible Development Board

1 2 3 4 5 6 7 8 9   LCR   21 20 19 18 17 16

ISBN   978-1-25-964479-5
MHID   1-25-964479-0

Sponsoring Editor
Michael McCabe

Editorial Supervisor
Stephen M. Smith

Production Supervisor
Lynn M. Messina

Acquisitions Coordinator
Lauren Rogers

Project Manager
Poonam Bisht,
MPS Limited

Copy Editor
Pallavi Sinha,
MPS Limited

Proofreader
Heather Mann

Indexer
Becky Hornyak

Art Director, Cover
Jeff Weeks

Illustration
MPS Limited

Composition
MPS Limited

CONTENTS

About the Author

Christopher Rush has a degree in computer science and has spent the last 10 years working for an electronics distribution company as a product manager for single-board computing. He is the author of *30 BeagleBone Black Projects for the Evil Genius*™ and *Programming the Photon: Getting Started with the Internet of Things*, both also published by McGraw-Hill Education.

PREFACE

This book is the perfect introduction to programming the Intel® Galileo development board. The Galileo was Intel's first development board aimed for the maker market and based around their own Intel Quark SoC. With built-in Internet of Things capabilities, this board unleashes the powerful hardware expanding its capabilities to Internet-connected hardware.

The Intel Galileo is fully compatible with the Arduino™-style programming language while also introducing its own libraries and features for connectivity. The board itself comes with the Arduino footprint for connecting Arduino-compatible hardware such as shields. On its own the board doesn't really do much, but once you learn how to connect hardware it is fully capable of acting as the brain of your projects, controlling things and sending data to cloud services using the on-board Linux.

This book presents you examples using the popular Grove system by Seeed-Studio, which allows you to interconnect hardware without the worries and frustration of dealing with circuits and soldering. Most of the examples use the parts commonly found in the Intel Galileo Grove Starter Kit.

The purpose of this book is to get you started with creating your own hardware projects with the Intel Galileo. You do not need any previous experience with circuits or programming, but general computer skills would be highly advantageous. *Programming the Intel® Galileo* is written to give you a wide variety of experience and a basic understanding of the capabilities of the Intel Galileo board. This book only covers the basics of how to program the board, the assumption being that you will then expand those skills to create your own exciting projects.

I would love to hear your thoughts and comments regarding this book, so I encourage you to contact me through www.rushmakes.com or on Twitter @rushmakes. You can download all the example code from the McGraw-Hill website www.mhprofessional.com/intelgalileo or through my GitHub, https://github .com/ChristopherRush/Programming-the-Intel-Galileo.

Christopher Rush

1

Introduction to the Intel Galileo

The Intel® Galileo Gen 2 is a board based on the Intel Quark System-on-Chip (SoC) X1000, a 32-bit Intel Pentium processor system, operating at speeds up to 400 MHz. This Quark system is capable of supporting the Yocto 1.4 Linux distribution that opens up further capabilities to the Intel Galileo board.

The board itself has a built-in Ethernet socket with additional support for Power over Ethernet (PoE), a Universal Serial Bus (USB) 2.0 host port for adding USB devices, a micro-SD slot for memory expansion, a mini PCI express (mPCIe) card slot, 20 digital input/output pins (six PWM outputs with 8/12-bit resolution and six analog inputs with 12-bit resolution), a micro-USB connection for USB client programming, an ICSP header, a JTag header, and two reset tactile buttons, all of which you can see in Figure 1.1.

The Intel Galileo Gen 2 board also features an integrated real-time clock (RTC), with an optional 3-V coin cell battery for operation between turn-on cycles of the board.

The Intel Galileo board also supports the use of Arduino™-compatible shields and can operate at either 3.3 or 5 V using a jumper pin header on the board. The board has been specifically designed to be hardware and software pin-compatible, with the Arduino shields based around the standard Uno R3 Arduino board. The shield headers have digital pins 0 to 13, analog inputs 0 to 5, an ICSP header, and universal asynchronous receiver/transmitter (UART) port pins that are all in the same location as those on the Arduino Uno R3.

Figure 1.1 *Intel Galileo Gen 2 hardware features.*

Hardware Summary

Microcontroller	SoC Quark X1000
Operating voltage	3.3 V/5 V
Input voltage	7–15 V
Digital I/O pins	14 (6 PWM outputs)
Analog input pins	6
Flash memory	512 kB
RAM	256-MB DDR3
SRAM	512 kB
Flash storage	8 MB
EEPROM	8 kB
Clock speed	400 MHz
Length	124 mm
Width	72 mm

Gen 1 and Gen 2 Comparison

Currently, two versions of the Intel Galileo are available: Generation 1 and Generation 2. The two boards that have been developed are considerably different:

- The Gen 1 Intel Galileo board does not have an on-board regulator, so the power supply has to be exactly 5 V. In contrast to this, the Gen 2 board has an on-board regulator, so it may be powered with any suitable power supply providing between 7 and 15 V direct current (DC). Some other properties of Gen 2 are as follows:

- 12 GPIOs fully native for greater speed and improved drive strength.
- 12-bit PWM for more precise control of servos and smoother response times.
- 12-V PoE capability.
- Serial console UART header is compatible with FTDI USB convertors.
- Console UART1 can be redirected to Arduino headers in sketches, which can eliminate the need for soft-serial.

On-Board Linux

The Yocto 1.4 Linux distribution is installed on the Intel Galileo, and you can easily access the various Linux functions with the system() call in the Arduino integrated development environment (IDE) program.

Power

The Intel Galileo Gen 2 can be powered only via an external power supply. The power adaptor can be connected by plugging a 2.1-mm center-positive plug into the board power jack. The board can operate on an external supply between 7 and 15 V DC. The power pins for the Intel board are as follows:

- VIN: The input voltage to the Intel board. You can access the voltage supplied via the power jack through this pin.
- 5 V: This pin outputs a regulated 5 V from the regulator board.
- 3.3 V: A 3.3-V supply is generated by the on-board regulator, which also provides the power supply to the Quark microcontroller.
- GND: Ground pins.
- IOREF: This pin on the Arduino board provides the voltage reference with which the microcontroller operates. This can be either 3.3 or 5 V, depending on the IOREF jumper position.
- 12-V PoE capability.

Buttons

There are two buttons on the Intel Galileo board with different functions:

- Reboot: It resets the Quark X1000 processor.
- Reset: It resets the Arduino sketch and also any additional attached shield.

Memory

The Quark X1000 microcontroller has 512 kB of embedded SRAM available. The board also has an additional 256 MB of DDR3 RAM and 8 MB of flash to store the firmware and any Arduino sketches. The on-board uSD slot supports uSD cards up to 32 G and can be used to provide the complete Yocto 1.4 Linux image.

Input and Output

The Intel Galileo has a number of input and output features, such as

- Serial: 0 (Rx) and 1 (Tx) pins, which are used to receive and transmit time to live (TTL) serial data to the Galileo board.
- Digital I/O: Digital pins 0 to 13 and analog pins A0 to A5 can all be used as a digital input or output, using the `pinMode()`, `digitalWrite()`, and `digitalRead()` functions in the Arduino IDE. All of these pins can operate at both 3.3- and 5-V logic.

NOTE   Each pin can provide or receive a current of 16 mA at 5 V or a current of 8 mA at 3.3 V.

- PWM: Pins 3, 5, 6, 9, 10, and 11 provide 8/12-bit PWM output with the `anlogWrite()` function in the Arduino IDE. The resolution of the PWM signal can easily be changed with the function `analogWriteResolution()`.
- SPI: SPI header or ICSP header on other Arduino boards has three pins that support SPI communication using the SPI Arduino library.
- Analog inputs: Pins A0 to A5 each have analog input that can provide 10/12 bits of resolution, which also can be changed using the `analogReadResolution()` function within the Arduino IDE.
- SDA and SCL: These support TWI communication using the wire library.

Communication

The Intel Galileo has a number of hardware features for communicating with a computer, another Galileo board, Arduino, or any other microcontroller board while also being able to communicate to devices such as phones and tablets. The board provides two UART controllers: UART 0 to Galileo headers 0 and 1, and UART 1 to 6-pin 3.3-V USB TTL FTDI header, optionally directed to Galileo headers 2 and 3.

The native USB port can also act as a USB host for connecting peripheral devices such as mice, keyboards, or smart phones. The on-board uSD card reader slot is accessible through the SD library functions. The communication between the Galileo and the SD card is provided by an integrated SD controller and does not require the use of the SPI interface like on other Arduino boards.

The on-board Ethernet interface is fully supported using the Ethernet library in the Arduino IDE. Like other Arduino devices and shields, it does not require the use of the SPI interface and library.

The Arduino software for the Intel Galileo includes a wire library to simplify the use of TWI/I2C bus.

The board provides an mPCIe slot that allows the full-size and half-size (with adaptor) mPCIe modules to be connected to the board and also provides an additional USB host port via the slot. Any standard mPCIe module can be connected and used to provide applications such as Wi-Fi or Bluetooth connectivity.

Programming

The Intel Galileo can be programmed with a special version of the Arduino software that has been written specifically for the Intel board. It is possible to make certain requests to the Linux Kernel with the `system()` calls within the IDE. This gives your Arduino sketches access to powerful utilities like Python, Node js, OpenCV, and all sorts of other applications.

NOTE   *The Intel Galileo forgets the sketch after powering down or when the board has rebooted. It is possible to boot the Galileo from the uSD card so as to restore the sketch from the same card.*

The default image of the Intel Galileo board comes flashed with Linux with libraries in the user space for integrating the Arduino wiring platform. Arduino on the Intel Galileo runs a little different from the usual Arduino microcontrollers.

Arduino runs in the Linux Kernel user space and is fully integrated with the IDE, which runs on your computer. The major benefit of using this method on the Galileo board is that developers can build native applications, install device drivers, change the Linux Kernel configuration, or even change the whole Linux distribution altogether.

At this stage you might be wondering why you decided to purchase the Intel Galileo board as opposed to other Arduino microcontrollers. With the Intel Galileo board, the Linux OS is responsible for handling all the digital and analog Arduino headers, which means that it avoids additional bridges that would normally need to be put in place and the Quark CPU does all the processing rather than a microcontroller. The seamless integration allows the Arduino APIs and Linux APIs to coherently exist and work together. This also means that in theory you can run multiple Arduino sketches from the command line in Linux using the SD card to store the sketches.

Summary

Looking at the hardware specification of the Intel Galileo, you see why it is a powerful board. This is one of the first maker's boards Intel brought to market and they have done a great job of putting the hardware together. The eradication of the need for a microcontroller is quite unique and offers something different to the market. In the next chapter, we will look at how to get connected to your board.

2

Getting Connected

In this chapter, we will look at the many ways in which you can connect your Intel Galileo board and get started straightaway creating Internet-connected hardware projects. Blinking a light-emitting diode (LED) is more commonly the first thing you will do when using a new hardware development board, and it is confirmation that everything has been set up correctly and is in perfect working order. For those of you who have programmed before, you will know that your first step is to print "Hello World." Getting an LED to blink on and off is our way of saying "Hello World." At the end of this chapter, you will learn the different hardware features of the Intel Galileo board and what you will need to work with it, and how to install the custom Arduino integrated development environment (IDE) program to upload the code to the board.

Power the Board

You can power up your Intel Galileo board in a number of ways, but by far the best way is using the standard direct current (DC) barrel jack providing 5 V DC. Alternatively, you can power up your board through Universal Serial Bus (USB), but this is not recommended due to the power limitations of USB, which cannot power both the board and power-hungry connected hardware. It is also important to note that you cannot power the board from anything other than 5 V; otherwise, you risk damaging the board and any hardware that is also connected to it.

NOTE   *If you are using an Intel Galileo Gen 1 board, make sure you power the board first before connecting the USB data cable.*

Figure 2.1 *Powering up the Intel Galileo board.*

Connect the 5-V-DC power adaptor to the DC socket on the Intel Galileo board as shown in Figure 2.1. Once plugged in and switched on at the main socket, the power LED should light up green.

Communicating with Your Board

The Intel Galileo has a number of ways to communicate with your computer, Arduino, or another microcontroller board. First and foremost, the Intel Galileo board supports universal asynchronous receiver/transmitter (UART) time to live (TTL) serial communication, which is available on digital pin 0 (Rx) and pin 1 (Tx). In addition to these pins, a second UART header provides RS-232 support and can be connected through a six-pin header as shown in Figure 2.2.

The USB client port allows for serial communication to a computer through USB. This provides a serial connection to the serial monitor or other application on your computer, and it is also used to upload sketches to the board through the Arduino IDE. This will be our preferred method of programming the Intel Galileo board and the one that we will show you later on in this book.

The USB host port should not be confused with the USB client port. The USB host port allows the Intel Galileo to act as a host for connected devices such as mice, keyboards, smart phones, and many more devices. You can connect a USB hub to this USB port to expand the number of ports up to 128 devices, if necessary.

Figure 2.2   *Connecting through UART pin headers.*

One of the more advanced ways of communicating with your Intel Galileo is to use the on-board mini PCI express (mPCIe) adaptor, which is a first for an Arduino device. This slot on the board allows both full-size and half-size (with adaptor) mPCIe modules to be connected easily to the Intel board and also can provide an additional USB host port. Any standard mPCIe module can be connected to provide application features such as Wi-Fi, Bluetooth, or cellular. Initially, the board provides support only for Wi-Fi using the mPCIe modules.

Getting connected to the Internet is important for projects involving the Internet of Things (IoT); that is why the Intel Galileo board comes with an RJ45 Ethernet adaptor built onto the board for fully wired networks. This is by far the easiest way to connect to your network and the Internet, and also the quickest. Unlike standard Arduino shields, the Ethernet adaptor is fully supported hardware and therefore does not require any SPI interface.

The on-board micro-SD card reader can be accessible through using the SD library, which we will cover later on in this book. The communication between the Intel Galileo and the SD card is provided by an integrated SD controller and does not require the use of SPI, unlike other SD Arduino shields and boards. The native SD interface runs at up to 50 MHz depending on the class of SD card being used.

Setting Up the Development Environment

NOTE   *Do not use the same power supply for Galileo Gen 1 and Gen 2. The Gen 1 board is rated at 3.3 and 5 V, whereas the Gen 2 is rated at 7 to 15 V.*

Now you should have a good understanding of both the Galileo boards and their hardware features; let us go ahead and set up the development environment. You will need to perform the following steps in order to get started:

1. First, connect the Intel Galileo board to your selected DC power supply through the barrel jack. Later on in the book you will learn how to connect the board through a standard PP3 battery connector, but for a stable power source and reliable results use a regulated power supply. When the board is plugged in and powered on at the main socket, you should see the LEDs on the board light up. Once the board is powered on, you can connect the USB client port to the USB port on your desktop or laptop computer as shown in Figure 2.3.

2. Now that you have your Intel Galileo board powered up, it is important to make sure that you have the latest firmware installed and up to date. Open up the Internet browser on your computer and go to the

Figure 2.3   *Intel Galileo Gen 2 board connected to the USB and power ports.*

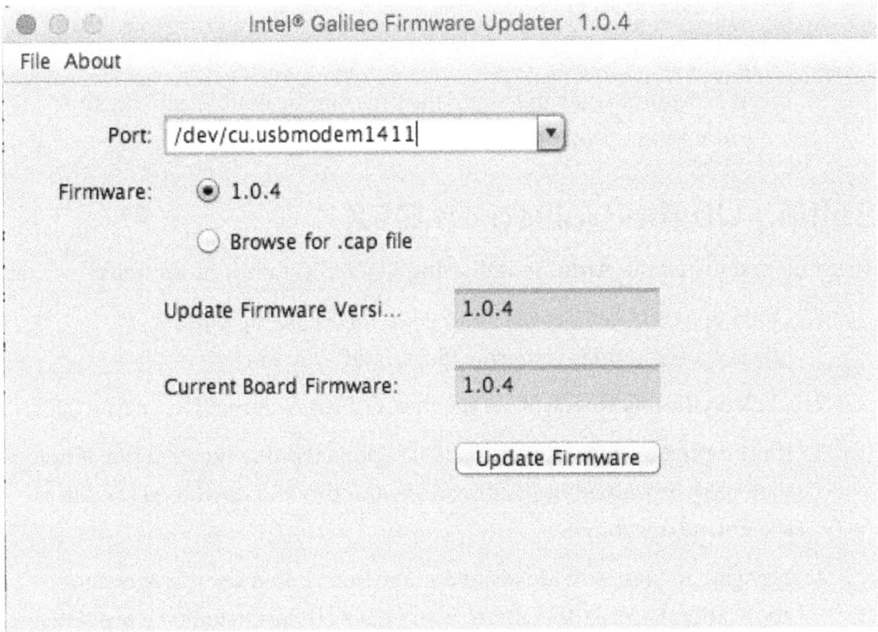

Figure 2.4 *Intel Galileo firmware update program.*

following URL: https://downloadcenter.intel.com/download/24748/. In the left-hand column, find your current operating system and download the firmware updater. Once it is downloaded, extract its contents using ZIP software and run the firmware program. Select the communications port in the drop-down box and click "Update Firmware" as shown in Figure 2.4. The process may take up to 5 minutes, so this is a good time to make some coffee before we proceed.

NOTE   Make sure that a stable power supply is used to power the Intel Galileo board.

Setting Up the Galileo on Windows

1. Next you need to download the Intel Galileo development environment, which is a custom version of the Arduino IDE developed by Intel:

   a. You can download the IDE from the following URL: https://communities.intel.com/docs/DOC-22226.

b. Open the file that you have downloaded by decompressing it using ZIP software, and locate the IDE program.

c. At this point you can launch the program by double clicking the Arduino.exe program file.

Setting Up the Galileo on OSX

To set up and install the Arduino IDE using MacOSX, follow these steps:

1. Open up your Web browser and go to the following URL: https://communities.intel.com/docs/DOC-22226.

2. Click on the link to download the Intel Galileo Arduino IDE on MacOSX.

3. If prompted by your browser, you can open the downloaded file; if not, then click download in your browser, and this will download the file to the downloads folder.

4. Navigate to your downloads folder and you should see the Arduino application here. At this point, if you have already installed a previous version of Arduino IDE for use with other Arduino boards, then you may want to rename the newly downloaded version to "Galileo" and keep both copies in the application folder; see Figure 2.5. Drag the downloaded IDE to your application folder.

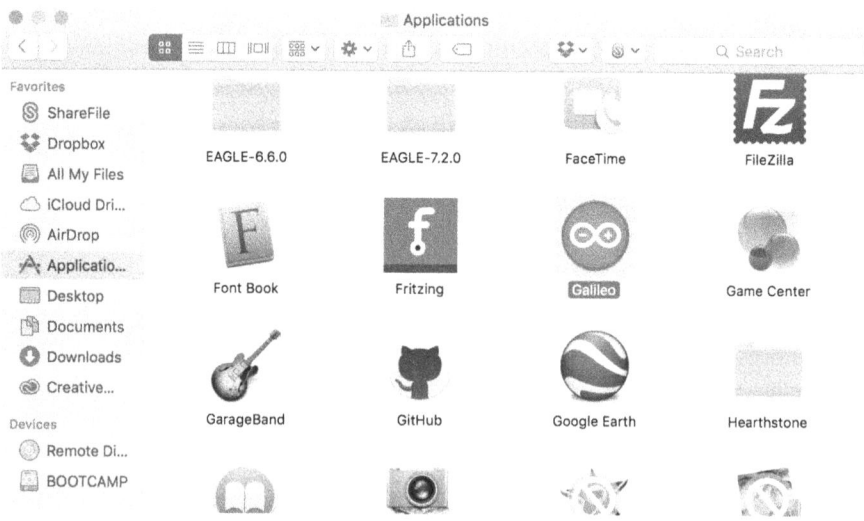

Figure 2.5 *Arduino MacOSX downloads folder.*

5. With your Galileo board powered through the DC barrel jack, connect the board from the client USB port to your computer or laptop, if you have not done so already.

6. Open up the applications folder in the Mac operating system and double click the Arduino (or Galileo if you have renamed it) to launch it.

Setting Up the Galileo on Linux

This set of instructions has been tested using Ubuntu; other types of Linux distributions may differ slightly. To set up the Galileo and install the Arduino IDE, follow these simple steps:

1. Open up your Web browser and navigate to the following URL: https://communities.intel.com/docs/DOC-22226.

2. Click on the link to download the Intel Galileo Arduino software for either Linux 64-bit or Linux 32-bit.

*NOTE   If you are unsure which system you are running your Linux on, you can type **uname -m** from the command line, which will return the system you are using; failing this, you can just run the 32-bit download as this also works on the 64-bit system.*

3. If prompted by your browser, save the .tgz file to your downloads folder directory.

4. Open up a terminal window by clicking its icon or using the shortcut Ctrl+Alt+T.

5. Change the current directory to that of where you downloaded the file to using the CD command followed by the directory location and then decompress the file using the tar program:

```
cd ~/Downloads
tar -xzf IntelArduino-1.6.0-Linux64.tgz -C ~/
```

6. Change the directory to where the extracted files have decompressed:

```
cd ~/Arduino-1.6.0
```

7. Now you can launch the Arduino IDE using the following command:

```
./Arduino
```

Troubleshooting Linux

Sometimes you may encounter some difficulties with the above steps. When launching the application, there is a possibility you may get an error regarding Java not being installed on the Linux distribution. You can install it with the following command in the terminal window:

```
sudo apt-get install default-jre
```

When the application has launched and the serial menu is grayed out, this could mean that when you launch the Arduino IDE you need to have root privileges to access the serial port. To do this, you can launch the program with the following command in terminal:

```
Sudo./Arduino
```

Hello World: Uploading Your First Code

Now that you have set up the board and installed the Arduino IDE, we need to check that everything is in working order. The best way to do this is to upload your first program to the board. At this point we are not interested in how things work or which part of the code does what; we just want to upload an example to blink an LED on and off to make sure that all the steps above have worked and the board is working as expected without any complications.

One of the good features about using the Arduino IDE is that it comes with a number of examples to get started with. Upload a basic example using the following steps:

1. Within the IDE, click File-Examples-01.Basics-Blink; see Figure 2.6.

2. When you click the blink example, a new sketch window will open up with some code in it. In the tool bar click the upload icon, which should start to compile all the code and then send to the board.

3. It may take some time to compile and upload to the board; you will also see some text in the console below the code, and you should see "Transfer complete."

When you click the upload button, the Arduino IDE compiles the sketch, which means it turned the code into a set of instructions that the Galileo board can

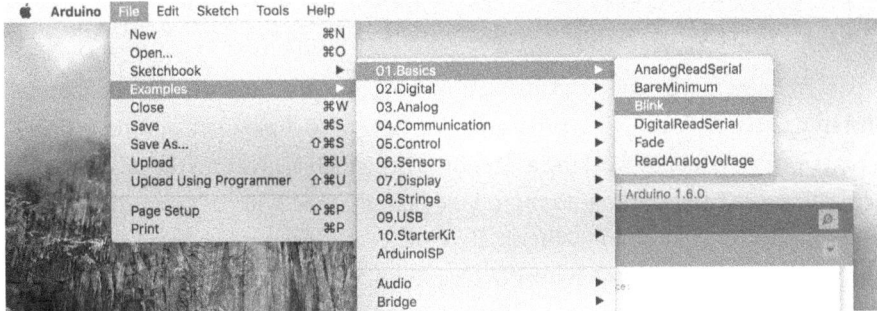

Figure 2.6 *Selecting the blink example in the Arduino IDE.*

Figure 2.7 *Blinking the on-board LED.*

understand. The Arduino IDE then uploads that program to the board, which is then run. Now when you look at the board you should see a green LED blinking on and off every other second. This LED is located next to the USB host socket and next to the power LED, which should be on; see Figure 2.7.

If you encounter any issues, you can always seek help from the Galileo support community at https://communities.intel.com/community/makers or the Arduino Forum at http://forum.arduino.cc.

Summary

In this chapter, we have looked at powering up the Intel Galileo board and installing the Arduino IDE program. We have also uploaded our first sketch program; everything is working as it should be, which is great. In the next chapter, we will look at the C programming language that we will be using to program some hardware later on in the book. Experienced programmers may want to skip the next chapter and head on over to Chapter 4, where we will look at controlling outputs.

3

Arduino-Style
C Programming

The language used to program the Intel Galileo is called C programming language. In this chapter, you will learn and understand some of the basic programming terms and concepts using this C programming language. You can apply what you will learn here to most of the sketches you will write throughout the book. To get the most out of your Intel Galileo, you will need to learn these basic programming fundamentals.

What Is Programming?

It may not seem obvious to a beginner what programming and a programming language are. When you look at the Intel Galileo programming code, you could probably hazard a guess as to what it is actually doing without any programming knowledge, but we need to look a bit further into how the programming language code goes from being lines of text to an action in real time, like turning an LED on or off.

When you click the upload button in the Arduino IDE, it starts a chain of events that results in your sketch being uploaded to the Intel Galileo board and being run. What it actually does is something called compilation, taking your lines of code as text and translating them into something called binary, which is a series of 1s and 0s that the Galileo's hardware will understand. In the previous chapter, you clicked the verify button before you actually flashed any of your code to the Galileo. This attempts to pre-compile the C code that you have written without

17

actually flashing it. Verifying your code also checks that what you have written makes sense in the C programming language. If you have written some code that is not within the C programming language, then when you verify your sketch it will return an error in the console.

Setup and Loop

The same goes when you try compiling a sketch with no written code at all; the error returned tells us that there is no `setup` or `loop` function in your code. These two functions of code are required and must always be present within your sketch. When you open up the Arduino IDE program, you should see a new sketch as shown in Figure 3.1. By default, when you open up a new sketch you will see the following code in the sketch:

```
void setup (){
}
void loop() {
}
```

When verifying the sketch, the compiler will tell you that it has successfully compiled your code and that everything was acceptable to the C language standard. At the bottom of the IDE in the console window, it will also tell you how much of the flash memory has been used on the device, in our case 80,230 bytes of a maximum of 10 million.

The Intel Galileo board has a total of 8 MB of internal flash memory for uploading your sketches with Intel's firmware; this is plenty of storage for even the largest of projects.

Let us take a closer a look at the setup and loop functions that will always be the starting point of every sketch that we write. We start off using the word `void` before `setup` and `loop` followed by a pair of curly braces. The line `void setup ()` means that we are defining a function called setup within our code. Some functions are already defined for us, such as `digitalWrite` and `delay`. `Setup` and `loop` are two functions that we must define for ourselves in every program we write.

We are not calling `setup` or `loop` like we do with `digitalWrite` or `delay`, but we are actually creating these functions so that the Intel Galileo itself can call them. This might sound a bit confusing, but the best way to think of it is that we are trying to shorten our code. Using void with both functions `setup` and `loop`

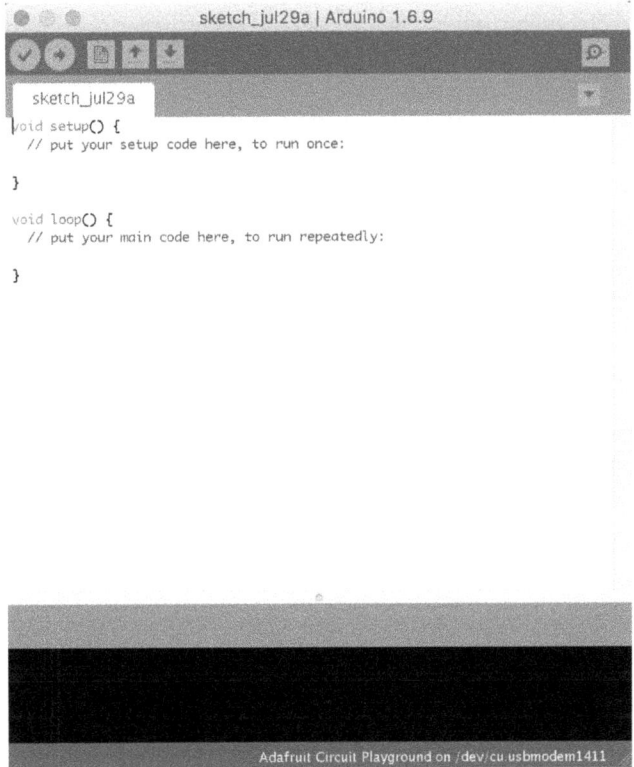

Figure 3.1 *Arduino sketch with default code.*

allows us not to return any values within the function unlike other functions do, so we have to say that these are void.

After the word `void` comes the function's name and the parentheses, which should contain any arguments. In our case, both the setup and loop do not contain any arguments, but we still have to include both the parentheses. Because we are defining a function within our code and not calling a function, we do not have to close it with a semicolon. Instead, we use the curly braces, where our code sits in between the function; this is known as the block of code. Just because we define the function's setup and loop, that does not necessarily mean we have to use those functions to hold any block of code; we simply need to define them in every sketch we write, although not using the function setup and loop in reality this may not ever happen.

Let us go back to our example sketch in Chapter 2:

```
void setup() {
    pinMode(13, OUTPUT);
}

void loop() {
    digitalWrite(LED, HIGH);
    delay(1000);
    digitalWrite(LED, LOW);
    delay(1000);
}
```

The `setup` function in our sketch uses one built-in function called `pinMode`. The function `pinMode` is used to set a particular pin on the Galileo to be either an input or an output. With this in mind, it is clear that we need to initially set our LED to be an output, which will in turn let us use the function `digitalWrite` later on. The `pinMode` is always used in the setup function because we only need to set the `pinMode` once in our sketch. The program would still work if you called this in the `loop` function, but for best coding practices it is always better to keep things that you call once in the `setup` function. Then you know where everything is that you only call once.

Variables

A variable is a place in the memory to store a piece of data. It has a name, a value, and a type. For example, the following statement defines the pin number:

```
int pin = 13;
```

This code creates a variable called pin whose value is 13, and its type is an integer that is written as `int` in the sketch. Later on in your code, you can refer to this variable by its name, at which point its value will be looked up and used. For example,

```
pinMode(pin, OUTPUT);
```

It is the value of the pin that will be passed into the `pinMode()` function. In this particular case, you do not actually need to use a variable; this statement would work just as well if you referenced the pin number straightaway like in the blink sketch example:

```
pinMode(13, OUTPUT);
```

The advantage of a variable in this case is that you only need to specify the actual number of the pin once, but another advantage is that you can use this more than once in your code. When naming your variable you can also use a descriptive name, which will make the significance of the variable clearer (e.g., a program controlling some LEDs might have variables called RedPin, GreenPin, etc.).

A variable has other advantages over a value such as a number. You can change the value of a variable in your code by using an assignment (indicated by an equals sign). For example,

```
pin = 11;
```

This will change the value of the variable to 11. You can see that we do not actually specify the type of variable when changing the value; only the name of the variable and the variable type will stay constant.

NOTE *Remember that you have to declare a variable before you can assign a value to it.*

You can of course make a copy of a variable within your code if needed. When you change the value of one, it does not affect the value in another. This is very useful for you when you change the value of a variable but you may also want to keep the original value handy if you need to go back to it. For example,

```
int pin = 13;
int pin2 = pin;
pin = 11;
```

Only the variable pin would have changed to 11, and pin2 would remain the same (13).

If you try changing the value of a variable before you have declared the variable, then you will receive an error in your code reading: "error: (variable name) was not declared in the scope." Scope refers to the part of your program in which the variable can be used. This is determined wherein you declare it; for example, if you want to be able to use your variable anywhere within your program, then you must declare the variable at the top of the program. This is known as a *global* variable. Following is an example of declaring a global variable:

```
int pin 13;
void setup() {
      pinMode(pin, OUTPUT);
```

```
{
void loop() {
      digitalWrite(pin, HIGH);
}
```

As you can see in the example, pin is used in both the `setup` and the `loop` functions. Both functions are referring to the same variable; therefore, it must be set as a global variable. If you only need to use a variable in a single function, then you declare it there, in which case its scope will be limited to that particular function. For example,

```
void setup() {
      int pin 13;
      pinMode(pin, OUTPUT);
      digitalWrite(pin, HIGH);
}
```

In this example, the variable pin can only be used inside the `setup` function. If you try to use it within the `loop` function, you will receive an error in your program. You may be wondering why we do not simply declare all variables as global variables at the start of the program. Having the variable inside the `setup` function makes it easier to find out what has happened to the value of the variable. When we use a global variable, the value can be changed anywhere within the whole program; this means you need to understand your program in order to know what happened to the variable. Sometimes it is much easier to debug when you only use that variable within its own scope.

Floats

All our examples so far have included int variables. Integer variables are by far the most commonly used type; however, there are others that you should be familiar with.

One type that will become more relevant later on in this book is a float. A good example is the conversion of temperature when using temperature sensors. The variable type is a number that may consist of a decimal place for more precise measurements such as 1.6. Take a look at the following formula:

$$f = c * 9/5 + 32$$

This is the formula for converting temperature from degrees Celsius to degrees Fahrenheit. If we give c the value of 23, then f will be 23 * 9/5 + 32 or 73.4. If we set f as a standard integer variable, then the value will return 73.

Take note of the order of the calculation. If we are not careful with the order of formulas, things may turn out differently when using integers. Take, for example,

$$f = (c/5) * 9 + 32$$

This formula would result in the following:

- 23 is divided by 5, which returns 4.6, which is then rounded down to 4.
- 4 is then multiplied by 9 and 32 is added to give a result of 68, which is way off our actual temperature value of 73.4.

For situations like this, we use floats. In the following example, our temperature conversion function is rewritten using floats:

```
float centToFaren (float c)
{
    float f = c * 9.0 / 5.0 + 32.0;
    return f;
}
```

We have also added .0 to the end of each value; this way our compiler will know that it should treat the values as floats and not as integers.

Boolean

Another common type of variable is a Boolean, in which values are logical. A Boolean has a value that is either true or false. The best example of Boolean logic is using the "if" statement; the condition set in an if statement can only be true or false:

```
int LEDpin = 5;

int switchPin = 13;

boolean running = false;

void setup()
{
   pinMode(LEDpin, OUTPUT);
   pinMode(switchPin, INPUT);
   digitalWrite(switchPin, HIGH);
}

void loop()
{
   if (digitalRead(switchPin) == LOW)
```

Data Type	Memory (bytes)	Range
Boolean	1	True or false (0 or 1)
Char	1	−128 to +128
Byte	1	0 to 255
INT	2	−32768 to +32767
Unsigned int	2	0 to 65536
Long	4	2,147,483,648 to 2,147,483,647
Unsigned long	4	0 to 4,294,967,295
Float	4	−34028235E+38 to +3.4028235E+38
Double	4	Same as float

Table 3.1 *Variable Data Types*

```
{
   delay(100);
   running = !running;
   digitalWrite(LEDpin, running)
}
}
```

Along with Boolean values, you can manipulate the values using Boolean operators. These are very similar when you perform arrhythmic calculations; the most commonly used Boolean operators are **and**, which is written as &&, and **or**, which is written as ||. In addition to these operators, there is also **not**, which is written using !. This value can be either "not true" or "not false." Table 3.1 shows all the data types that are available and which format of values should be used.

Another thing to remember is what happens when your values exceed the limits. This causes odd things; for example, if you have the byte variable with a value of 255 and you suddenly add a 1 to this value, then it returns a zero. Similar to this, if you add a 1 to an integer with the maximum value of 32,767, it becomes a negative value of −32,768. Usually you can get away with most of your data type being an integer, so sometimes it is best to go along with this.

Char

The data type char is a byte that represents an ASCII (American Standard Code for Information Interchange) character. ASCII is a system from the very early days of computing for translating between bytes and characters. A char typically takes only up to 1 byte of memory that stores a character value. Single characters are written in single quotes—'A'—and multiple characters are written in double quotations—"ABC." In theory, a char is stored as a number according to the ASCII table (e.g., A is equal to the number 65). Here is how you create and assign a char variable:

```
char letter = 'A';
char letter = 65;
```

Both examples are correct.

Commands

The C language on the Galileo has a number of built-in commands. In this section, we will explore some of these commands and see how they can be used in your sketch.

If Statement

In our examples so far, we have assumed that your lines of programming will be executed in order, one after the other. But what if we want to execute a block of code when something happens in the code. For this we can use an if statement, which is used in conjunction with a comparison operator, and test whether a certain condition has been reached, such as an input being higher or lower than a certain number. The formatting for an if statement is as follows:

```
if (variable > 50)
{
//Write your code here
}
```

The program tests to see if the variable value is greater than 50. If it is, then the program takes a particular action and executes the code between the curly braces. If the condition is not reached, then the program skips over these sections and moves on to the next lines of code. Also, the curly brackets may also be omitted after an if statement; if this is done, then it is the next line of code that becomes the only conditional statement as shown in the next example:

```
if (x > 50) digitalWrite(LEDpin, HIGH);
if (x > 50)
digitalWrite(LEDpin, HIGH);
if (x > 50) { digitalWrite(LEDpin, HIGH); }
```

Operator	Meaning	Examples	Result
<	Less than	9 < 10	True
		9 < 9	False
>	Greater than	10 > 9	True
		10 > 10	False
<=	Less than or equal to	10 <= 10	True
		9 <= 10	True
>=	Greater than or equal to	10 >= 10	True
		10 >= 9	True
==	Equal to	9 == 9	True
!=	Not equal to	9 != 9	False

Table 3.2 *Comparison Operators*

```
if (x > 50) {
     digitalWrite(ledPin1, HIGH);
     digitalWrite(ledPin2, HIGH);
}
```

All of the examples are correct for using an if statement. In the example we use the symbol >, which means more than. This is a comparison operator. These operators can be found in Table 3.2.

Just remember that when using the equal sign, you must use a double equals symbol, which is a *comparison operator* rather than a single equals symbol, which is an *assignment operator*. It is easy to get the two mixed up when using conditions. If we accidentally used a single equals operator, then the if statement would always return a true condition. This is because the C language always evaluates the statement as an assignment, so in our example x would be set to 50 and would always be true.

There is another form to the if statement, if the condition has not been reached or it is false. We will use this in some practical examples later in the book.

For Loops

You may find yourself also wanting to run a series of commands a number of times within your programs. We know already that we can use the loop function; when all the lines of code in the loop function have been run, it will just start over at the beginning of the loop. This is great, but sometimes we only want to run our code a few times to get a certain desired result and then stop. For this we can use something called a "for" statement, which is used to repeat a block of code within the curly braces. Usually you would use an increment counter to determine how many times you want to loop the code. The "for loop" is very useful for any kind of repetitive operation and is often used in combination with arrays. There are three main parts to the for loop header:

```
for (initialization; condition; increment) {
      statements
}
for (int x = 0; x < 10; x++) {
      delay(500);
}
```

The *initialization* happens first and only once, each time through the loop the *condition* is tested; if it is true, then the statements are executed, as well as the *increment*, and then the condition is tested again. The condition becomes false, when the loop is repeated more than 10 times.

While Loops

Another useful way of looping in C is to use the while command in place of the for command. A "while loop" will loop continuously and indefinitely until the expression inside the parentheses becomes false. Something has to change the tested variable; otherwise, the while loop will never end and will be stuck in an infinite loop. The syntax for using a while loop is as follows:

```
while (expression) {
      statements
}
int i = 0;
while (i = < 10)
{
      delay(500);
      i ++;
}
```

The expression in the parentheses after the while must be true to stay in the loop. When it is no longer true, then the sketch continues running the commands after the curly braces. You may also notice the following line:

i ++;

This is just C shorthand for the following expression:

i = i + 1;

Arrays

Arrays are a way of containing a list of different values. Unlike what we have learned before where variables have contained only a single value, usually by an int data type, an array contains a list of values. You can easily access any one of those values by its particular position in the array. In most programming languages, and in fact computer science in general, the first value is always represented as a 0 rather than a 1; this means that the first variable is actually element zero. Here are some following examples of how to declare an array in your code:

```
int myValue[6];
int myPins[] = {2, 4, 8, 3, 6};
int mySensVals[6] = {2, 4, -8, 3, 2};
char message[6] = "hello";
```

You can declare an array without initializing it, as in myValue. In myPins, we declare an array without explicitly choosing a size. The compiler counts the elements and creates an array of the appropriate size. Finally, you can both initialize and size your array, as in mySensVals.

NOTE *When declaring an array of type char, one more element than your initialization is required to hold the required null character.*

Arrays are zero indexed. So referring to the array initialization above, the first element of the array is at index 0, hence mySensVals[0] == 2, mySensVals[1] == 4, and so forth.

It also means that in an array with 10 elements, index nine is the last element, hence:

```
int myArray[10]={9,3,2,4,3,2,7,8,9,11};
      // myArray[9]      contains 11
```

// myArray[10] is invalid and contains random information (other memory address). For this reason, you should be careful when accessing arrays. Accessing past the end of an array (using an index number greater than your declared array size −1) is reading from memory that is in use for other purposes. Reading from these locations is probably not going to do much except yield invalid data. Writing to random memory locations is definitely a bad idea and can often lead to undesirable results such as crashes or program malfunction. This can also be a difficult bug to track down.

Strings

A string is a sequence of characters and a way for your Intel Galileo to deal with text. It is highly unlikely that we would use strings within our code; however, if you are using an LCD, then string might come into play.

All of the following are valid declarations for strings:

```
char Str1[15];
char Str2[8] = {'a', 'r', 'd', 'u', 'i', 'n', 'o'};
char Str3[8] = {'a', 'r', 'd', 'u', 'i', 'n', 'o', '\0'};
char Str4[ ] = "arduino";
char Str5[8] = "arduino";
char Str6[15] = "arduino";
```

Possibilities for declaring strings:

- Declare an array of chars without initializing it as in Str1.

- Declare an array of chars (with one extra char), and the compiler will add the required null character, as in Str2.

- Explicitly add the null character, Str3.

- Initialize with a string constant in quotation marks; the compiler will size the array to fit the string constant and a terminating null character, Str4.

- Initialize the array with an explicit size and string constant, Str5.

- Initialize the array, leaving extra space for a larger string, Str6.

Generally, strings are terminated with a null character ASCII code 0. This allows functions to tell where the end of a string is. Otherwise, they would continue reading subsequent bytes of memory that are not actually part of the string.

This means that your string needs to have space for one more character than the text you want it to contain. That is why Str2 and Str5 need to have eight characters, even though it is only seven—the last position is automatically filled with a null character. Str4 will be automatically sized to eight characters, one for the extra null. In Str3, we have explicitly included the null character (written '\0') ourselves.

It is possible to have a string without a final null character (e.g., if you had specified the length of Str2 as seven instead of eight). This will break most functions that use strings, so you should not do it intentionally. If you notice something behaving strangely (operating on characters not in the string), this could be the problem.

Coding Best Practices

The Arduino compiler does not pay any attention to how you layout your code, but it does require that you write all your code on a single line with a semicolon between each statement. If you think about how you may read a book, usually the formatting is very similar—you have your table of contents, chapters, paragraphs, and indexing.

Formatting your code is a personal choice—some like to keep it messy and some like to keep things neat and tidy with additional commenting between sections. Usually keeping code tidy is best; it helps debug code much quicker and also allows someone else to easily read your code if you collaborate with others.

Indentation

You can see in some of the example sketches that we always use some sort of indentation for the code in the left margin. Indentation is usually determined by the curly brackets and forms a hierarchical structure to the whole code. Using the example below, we can see that we have void loop() as our top level and within this we have a small amount of code followed by another sublevel if.

```
Void loop()
{
        int count = 0;
        count = ++;
        if   (count == 10)
            {
                    count = 0;
                    delay(1000);
            }
}
```

If we add another if statement within the first if, then we would simply increase the indentation by a further 1 or 2. To indent from the left margin, you can simply press Tab on the keyboard. You might find this a bit tedious, but when reviewing your code later, it will become apparent how useful this is.

Commenting Your Code

Comments in your code are text that the compiler does not read and simply ignores. The purpose of a comment may be to provide some additional information to either you the programmer or someone who is reading your code. If your program has a lot of code divided into many sections, then you can also use a comment as a title or header into that section. This can be useful when debugging code, so you do not need to search for the section you need to edit or change. There are currently two forms of syntax in which you can write comments in your code:

- The single-line comment, probably the most common one that you will use, starts with two backslashes together //. When using a single-line comment, you cannot carry on with code on the same line. If this happens, the code will be ignored by the compiler as it thinks it is part of the comment.

- The multiline comment is separated by /* and */. You can use this at the start of your program to introduce the code and write a short description of what your program does.

The following example shows both types of commenting syntax:

```
/* This is an example of how to use different types of
comments with your program.
Written by Christopher Rush
*/
void loop() {
        int count = 0;
        count ++; //adds plus one to the integer count
        if (count == 10) {
                count = 0;
                delay(1000) //pauses for 1 second
        }
}
```

In this book, I will stick to using single-line comments, usually to help explain what is happening in the code. This is useful if other people are going to use the code or snippets of the code in their projects. Sometimes knowing when and when not to use comments can be confusing for beginners; however, I usually follow a couple of simple rules that should make things a bit easier. Comments should be used to:

- Explain anything that can be a little tricky to comprehend.

- Describe something the user may need to do that is not written in the code; for example, //**LED must be wire up to pin D1.**

- Leave notes or instructions for yourself; for example, // **Note: tidy up this code with an easy function.**

The last point can be very useful to comment using either Note or Todo, which reminds you that you need to come back to this point sometime in the future. Some IDE compilers allow you to search for keywords.

Whitespaces

The compiler program will always ignore any whitespace lines in your program unless they are spaces that separate words in your code. The following code, for example, will still work, but reading it or debugging would prove to be very difficult:

```
void loop() {int
count = 0; count ++; if(
count==10) {count =0;
delay(1000);}}
```

Some users tend to put space between everything, and others try to create a more streamlined format. It does not matter either way; the compiler will still read the code in the way it is intended to.

```
int count = 0;
int count=0;
```

Summary

You may just want to get down to the nitty-gritty of the Intel Galileo, but it is important to understand the basic concepts of programming the board. In this theoretical chapter, you have learned how to write your sketches so that they are easy to read and understand as well as adopting the Arduino style of programming language, which will save you a lot of time and effort when writing your own sketches for the Galileo.

In the next chapter, we will look at programming some electronics using some output devices such as LEDs, motors, and servos for both analog and digital circuits.

4

Programming Outputs

This chapter introduces you to controlling output devices such as light-emitting diodes (LEDs), relays, and buzzers. Output devices are usually used to communicate information such as the status of a circuit or the ability to switch something on or off like a motor or a servo device. The Intel Galileo is all about connecting physical devices to the world, and this means connecting electrical components to your Intel Galileo board.

The outputs on the Intel Galileo are digital, which means switching between 0 and 3.3 V or 0 and 5 V. Outputs can also be analog signals, which allow you to set a varying voltage to a device of any value between 0 and 3.3 V or 0 and 5 V

Figure 4.1 *SeeedStudio base shield.*

depending on the board configuration. This book is primarily about software programming and teaching you the concepts of the programming language to create your very own projects. Therefore, we won't get too engaged with the hardware side of things and the complexity of the circuit but rather focus on the programming. With this in mind, I will be using the SeeedStudio Grove modules. The Grove modules require no soldering; they are plug-and-play building block–type modules that connect to the Galileo board through an interface board called a "base shield" as shown in Figure 4.1.

Experiment 1: Understanding Digital Outputs

The Intel Galileo has a whole host of pins available from D0 to D13 and A0 to A7. All these pins, by default, are output pins, but we can configure them in such a way in out sketches that they become digital or analog output pins that can control output devices connected to the board.

To understand a bit about how digital output pins work, we can conduct a simple experiment. This experiment involves the use of a basic digital multimeter and some prototyping wire as listed in Table 4.1.

For this experiment, we will use digital pin 0. Insert one piece of prototyping wire into the pin header for digital pin 0 and connect another piece of prototyping wire into the pin labeled GND, which is for ground voltage. You can see the arrangement in Figure 4.2.

Take your digital multimeter (if your multimeter has crocodile clips then you can clip the crocodile clips to the bare ends of the prototyping wire) and connect the pin D0 to the positive probe of the multimeter and the GND pin to the negative probe of the multimeter to create a circuit to the multimeter. If your multimeter is not supplied with crocodile clips, then you can simply wrap the prototyping wire around the end of the probe and affix with some electrical tape to make a good contact; see Figure 4.3.

Description	Appendix
Intel Galileo board	M1
Jumper/prototyping wire	H1
Digital multimeter	H2

Table 4.1 *Components and Hardware for Experiment 1*

Figure 4.2   *Prototyping wire inserted into pins D0 and GND.*

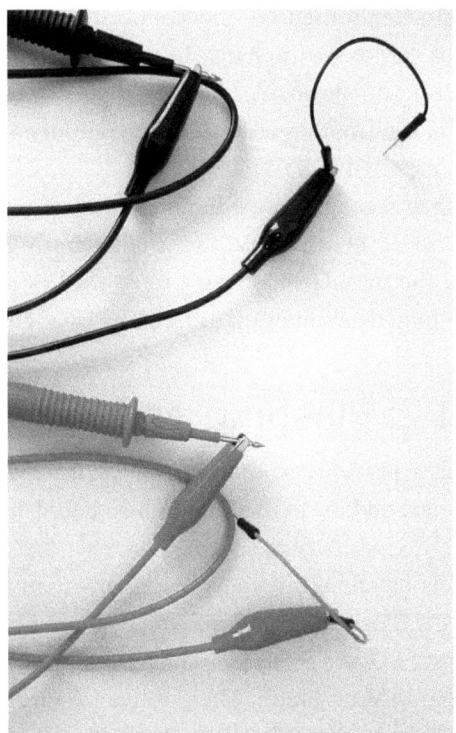

Figure 4.3   *Multimeter probes connected to the prototyping wire.*

The next step is to turn on the multimeter and set the range somewhere between 0 and 20 V DC or 0 and 10 V DC, as we already know that the maximum voltage range on the digital pin on the Galileo is between 0 and 5 V. Now let us load the following sketch to the Intel Galileo board and see what happens.

Sketch 1—Digital Outputs

```
int pin = 0;

void setup() {
pinMode(pin, OUTPUT);
}

void loop() {
digitalWrite(pin, HIGH);
delay(1000);
digitalWrite(pin, LOW);
delay(1000);
}
```

So with the multimeter turned on and connected to the pins on the Intel Galileo, you should be able to see its change in readings from 0 to 5 V every other second as the sketch runs on the board. This is a simple way to explain how digital pins work on the Intel Galileo; they send a voltage output value of either high or low, which is then interpreted into 0 and 5 V.

This is accomplished through the Arduino commands `digitalWrite` and setting the argument to `HIGH` or `LOW`. With the following in mind, you can easily see how we can use the Intel Galileo board to control electronics devices and components by switching them on or off using some basic circuitry.

Experiment 2: Flashing an LED

Progressing on from the previous experiment, we will use the same principle, but this time we will connect an LED to the Intel Galileo board and flash it on and off every other second.

LEDs will definitely be one of the most commonly used parts when creating your very own projects. They are inexpensive and easy to obtain from your local electronics store. Most LEDs are polarized, which means that it does matter which way you connect them to your circuit. The positive leg on the LED is called the anode, and the negative leg on the LED is the cathode. If you look at your LED on the top of the plastic shell, you can usually identify the flat side to the casing at the rim; this side is the cathode. Another simple way to determine which side is anode and which is cathode is by looking at the lengths of the LED legs. The longest leg is always the anode and the shortest is the cathode, as shown in Figure 4.4.

With all different types of LEDs, the current only flows in one direction: from the anode to the cathode. As a result, the anode should always be connected to the power source. In our instance, this will be the voltage output from the digital pin on the Intel Galileo board, which is 5 V. It is also common to see LEDs run in series with a resistor. Resistors are not polarized; therefore, you do not need to worry about how they are connected to the circuit. Table 4.2 shows the components and hardware that we will be using for this experiment and Figure 4.5 shows the schematic reference.

Figure 4.4 *5-mm LED.*

Schematic Reference	Description	Appendix Reference
M1	Intel Galileo	M1
	Grove base shield	H3
L1/R1	Grove LED module	D1

Table 4.2 *Components and Hardware*

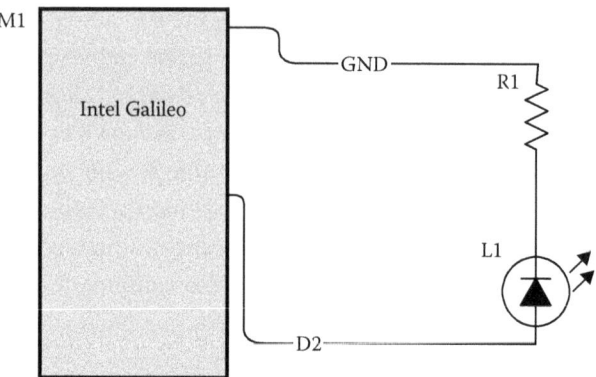

Figure 4.5 *Experiment 2 schematic diagram.*

Figure 4.6 *Inserting the base shield into the Galileo board.*

You will need to connect the Grove base shield to the Intel Galileo by inserting the shield pins into the Galileo's Arduino-compatible pin headers as shown in Figure 4.6.

For this experiment, we are going to be using digital pin 2 on the Intel Galileo board. Carefully selecting this pin also frees up pins 0 and 1 if you decide to use UART connectivity later on. The Grove LED module shown in Figure 4.7 already

Figure 4.7 *Grove LED module connected to the base shield.*

has a resistor in series with the LED, which will act as a current limiter for our LED. Connect the Grove LED module to the D2 grove connector on the base shield using the four pin male-to-male cable.

For this basic experiment, we are going to use the same program we used when we tested the voltage outputs of the Intel Galileo using the digitalWrite function to set the output to either HIGH or LOW.

Sketch 2—Flashing an LED

```
const int pin = 2;

void setup() {
pinMode(pin, OUTPUT);
}

void loop() {
digitalWrite(pin, HIGH);
delay(1000);
digitalWrite(pin, LOW);
delay(1000);
}
```

To load this sketch, verify the code that pre-compiles it and then click upload, which will then send the sketch to your Intel board and run the program on the

Galileo. You should see the LED flash on and off every other second as instructed by our code using the `digitalWrite` function. This is a simple example program that shows the basics of using physical outputs on the Intel Galileo board.

Experiment 3: LCD

One of the many benefits of using development boards is that they can operate independently from any computer. But if you want to display information about the status of certain connected hardware, then an LCD backlight is the perfect companion. An LCD is a common component on most consumer electrical goods, such as DVD players or Hi-Fi systems, to display track information. We can use this LCD module to show the status of sensor values, timing information, settings or progress bars, and much more. In this experiment, we will learn how to connect the Grove LCD module to the Intel Galileo and display some basic text on the screen.

To complete this example, we will use the Grove LCD backlight module as shown in Figure 4.8. This will save you some time and allow you to concentrate more on the software side of things.

The LCD Grove module enables you to set the backlight color to whichever you like using red, green, blue (RGB) values. The Grove module uses I2C as the

Figure 4.8 *Grove LCD backlight module.*

communication method with the Intel Galileo board. The advantage of this is that it significantly reduces the number of I/O pins that would be required from about 10 to just two pins. The LCD we are using is more commonly called a 16 × 2 character display; the configuration gives us a total of two rows with up to 16 different characters on each row, 32 in total. There is plenty of space for us to display information such as date and time with ease.

For this experiment you can use any of the I2C ports on the base shield—there are four in total. The hardware requirements for this experiment can be found in Table 4.3.

Connect the Grove LCD to the base shield using any of the available I2C sockets on the shield as shown in Figure 4.9.

Before we start programming the LCD, there are a couple of things we need, such as the LCD libraries from SeeedStudio. Libraries are a very efficient way of extending the Arduino environment. Libraries provide extra functionality for use

Description	Appendix Reference
Intel Galileo	M1
Grove base shield	H3
Grove LCD module	H4

Table 4.3 *Components and Hardware*

Figure 4.9 *Grove LCD character connected to the base shield.*

with your sketches. A number of libraries come pre-installed with the Arduino IDE, such as:

- EEPROM

- Ethernet

- GSM

- LiquidCrystal

- SD

- Servo

- SPI

- Wi-Fi

- Wire

In our sketch, we are going to use the library wire and the Seeed RGB LCD library for accessing the LCD functions. In order to import the library into our sketch, we must first download it from the Seeed GitHub webpage, https://github .com/Seeed-Studio/Grove_LCD_RGB_Backlight. Download the zip file and extract its contents. In the Arduino IDE, add the library to the list of imported libraries by navigating to *sketch > import library > add library*; then select the folder where you download and extracted the RGB LCD library. Go back to the library list *sketch > import library*, and you should now see the RGB LCD library in the list. Simply click this to add it to your sketch.

Sketch 3—LCD

```
#include <Wire.h>
#include "rgb_lcd.h"

rgb_lcd lcd;

const int colorR = 100;
const int colorG = 0;
const int colorB = 100;


void setup() {

    lcd.begin(16, 2);
```

```
    lcd.setRGB(colorR, colorG, colorB);

    lcd.clear();

    lcd.setCursor(0, 0);
    lcd.print("Getting Started");

    lcd.setCursor(0,1);
    lcd.print("Intel Galileo");
}

void loop() {

}
```

Let us take a closer look at the sketch by breaking down the code in sections. The first part of the sketch is where we import our libraries as previously mentioned. As the LCD module uses the I2C communication, we need to include the wire library to access this. As mentioned before, we use the RGB LCD library from Seeed, which gives us access to all the functions for the module.

Next we rename the LCD library to something a little easier:

```
rgb_lcd lcd;
```

The next set of variables are the RGB values of the backlight LED of the LCD. These values range from 0 to 255, and you can easily mix the values to create lots of different shades. I have set these values as constant integers because I do not want to change the values at any other point during the course of the program; however, you could do so by listing the variables as standard integer values. For example, you may want to change the color to red if something needs to grab your attention or green to inform a user that the system status is OK.

```
const int colorR = 100;
const int colorG = 0;
const int colorB = 100;
```

In the setup part of our sketch, we must first call the initialization command to set up a 16-character, two-row LCD.

```
lcd.begin(16, 2);
```

After we initialize the LCD, we can start to issue commands to it. First, we set the backlight color using the RGB values in our variables.

```
lcd.setRGB(colorR, colorG, colorB);
```

Now that everything is set up, we can write our first text to the LCD. First make sure the cursor is set to display the text. The first text should begin on the first line starting from the character farthest to the left, which is column 0 and row 0 (in Arduino, always use 0 as the first number); to set this we use the function setCursor. To print the text, we want to display, we simply use the function print followed by the text; remember that we are limited to 16 characters in total — the text will simply not run onto the next row below unless we tell it to.

```
lcd.setCursor(0, 0);
lcd.print("Getting Started");
lcd.setCursor(0,1);
lcd.print("Intel Galileo");
```

When you load the sketch, you should see the LCD light up with text on the display on both lines. There are also other features that you can use to accomplish this, such as scrolling text; this is particularly useful when you want to display more than 16 characters on the screen. We can accomplish this with the command lcd.scrollDisplayLeft. Let us take a look at the next sketch:

Sketch 4—LCD Scrolling Text

```
#include <Wire.h>
#include "rgb_lcd.h"

rgb_lcd lcd;

void setup() {
    lcd.begin(16, 2);
}

void loop() {
    lcd.clear();

    lcd.print("Getting started with the Intel Galileo");

    delay(500);

    for (int positionCounter = 0; positionCounter < 24;
positionCounter++) {
```

```
    lcd.scrollDisplayLeft();

    delay(150);
  }
  delay(1000);
}
```

You can see in the loop that there is a for loop that sets the condition for scrolling the text. If you remember from Chapter 3, the for loop has three sets of conditions: initialization, condition, and increment. In the condition, we set the amount of characters that we want to scroll left or right and we increment the integer value by 1 each time the for loop has run.

Experiment 4: Switching High-Voltage Appliances Using a Relay

One of the great limitations of most development boards is the ability to switch mains voltage appliances. However, with a small relay module, which is capable of switching 250-V 10-A mains electric, you can easily control the switching state.

Relays are very useful for home automation projects to control your internal house lighting, garage door, even your coffee machine or toaster—anything that switches on or off using mains voltage. For this experiment, we will use the Grove relay module, which is based on a solid-state relay.

A solid-state relay is a switching device that, when a small voltage is applied through the control pin, switches and creates a circuit for the input voltage that is much higher. Solid-state relays are much preferred because of their fast switching ability, and unlike other relays, they do not wear out as there are no mechanical parts.

The components required for this experiment can be found in Table 4.4 and the schematic reference in Figure 4.10. The actual wiring of the circuit is really simple: the live voltage from the power source connects to the lamps voltage in and the ground connect from the lamp to the relay; the relay connects back to ground from the power source to complete the circuit.

In this experiment, we are going to program the relay to switch on and off periodically using some simple coding techniques. We will set the state of the relay whenever we switch the relay on and off; this makes it easier to determine the current state of the relay and whether it needs switching HIGH or LOW.

Schematic Reference	Description	Appendix
M1	Intel Galileo	M1
	Grove base shield	H3
RY1	Grove relay	S1
LS1	Low-voltage lamp	D2
V1	9-V battery	H5

Table 4.4 *Components and Hardware*

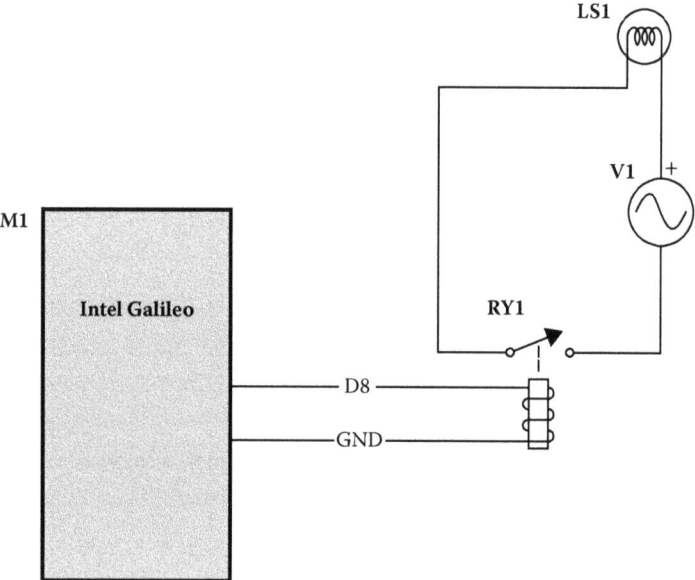

Figure 4.10 *Experiment 4 schematic diagram.*

Sketch 5—Relay Switching

```
const int relayPin =  8;
int relayState = 0;

void setup()
{

    pinMode(relayPin, OUTPUT);

}

void loop()
{
```

```
if (relayState == 0)
{
    digitalWrite(relayPin, HIGH);
    relayState != relayState;
}
else if
{
    digitalWrite(relayPin, LOW);
    relayState != relayState;
}
delay(2000);
}
```

Let us take a closer look at the sketch in detail. The first set of variables declare the pin number, which is the digital pin to which the relay is connecting; this is a constant integer value as the pin number never changes. The relay state integer is the state of the relay, whether HIGH (1) or LOW (0); this variable changes when we switch the relay on and off.

```
const int relayPin =   8;
int relayState = 0;
```

Because the relay is a switching device, we set the pin to an output pin in the setup function.

```
pinMode(relayPin, OUTPUT);
```

The most interesting part comes in the loop function; we check the state of the relay before we do anything else. By default the relay will be in an off state as set in the global variable. We control the relay in an if statement based upon the state variable; if the relay state is 0 then we switch the relay to HIGH and change the relay state to 1 and vice versa in the else statement if the relay is in an on state.

```
if (relayState == 0)
 {
    digitalWrite(relayPin, HIGH);
    relayState != relayState;
}
 else if
 {
    digitalWrite(relayPin, LOW);
    relayState != relayState;
}
```

After we switch the relay, we delay the loop for 2 seconds before we switch the state back again.

```
delay(2000);
```

Load the sketch to your Intel Galileo board and you should see the lamp switching on and off every few seconds or so. This is a great example of how to use relays in your projects; combined with inputs in the next chapter, you can easily turn on the lights when it gets dark using a light-level sensor or similar.

Experiment 5: Controlling a Servo Motor

DC motors are excellent drive motors, but they are not really good for precision work because there is no feedback. Without using some sort of encoder device, you will never know the exact position of a DC motor. Servo motors, or servos, are unique in that you can command them to rotate to a particular position and stay there until you tell them otherwise. A good example of a servo motor is an actuating door lock.

When using servos, there are a couple of different types to consider, such as standard and continuous rotation. Standard servos operate from 0 to 180 degrees. Servo control is achieved by sending a pulse of a particular length. The length of the pulse determines the absolute position to which the servo will rotate. This is due to a small potentiometer in the servo that measures its position; when you remove the potentiometer, it becomes free to continuously rotate.

Unlike a standard DC motor, servo motors have three pins: power (red), ground (black or brown), and signal (white or orange). The wires are color coded and typically in order; they look like the ones shown in Figure 4.11.

Also in contrast to standard DC motors, servos have a dedicated control pin that tells the servo which position to turn to. The power and ground lines of a servo should always be connected to a power source. Servos are controlled using adjustable pulse widths on the signal line. For a standard servo, sending a 1-ms 5-V pulse turns the motor to 0 degrees and sending a 2-ms 5-V pulse turns the motor to 180 degrees. Once a pulse has been sent to a servo, it then turns to that position and will remain there until instructed to move to another position by another pulse signal. If you want the servo to hold its position—that is, resist any movement—then you need to resend the same command every 20 ms or so.

Figure 4.11 *Servo motor.*

Schematic Reference	Description	Appendix
M1	Intel Galileo	M1
	Grove base shield	H3
S1	Grove servo motor	H6

Table 4.5 *Components and Hardware*

In this experiment, we will use a standard 0–180 servo motor and we will pan it from 0 to 180 and back to 0. The hardware for this experiment can be found in Table 4.5 and the schematic diagram in Figure 4.12.

Sketch 6—Servo Sweeping

```
#include <Servo.h>

Servo myservo;

const int pin Servo = 3;

int pos = 0;
```

```
void setup()
{
  myservo.attach(pinServo);
}

void loop()
{
  for(pos = 0; pos < 180; pos += 1)
  {
    myservo.write(pos);
    delay(15);
  }
  for(pos = 180; pos>=1; pos-=1)
  {
    myservo.write(pos);
    delay(15);
  }
}
```

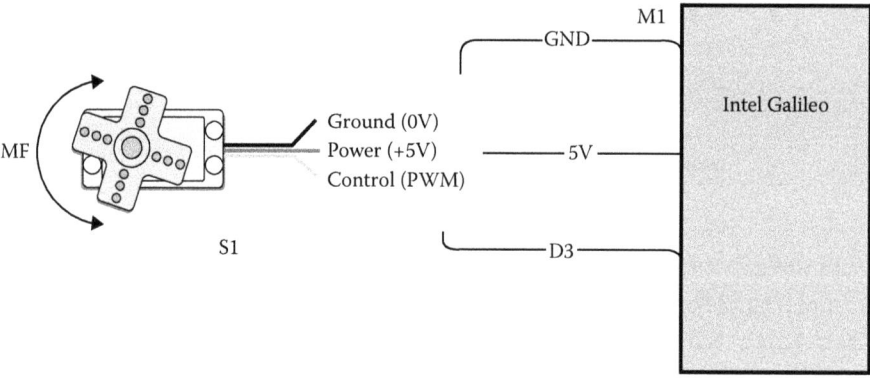

Figure 4.12 *Experiment 5 schematic diagram.*

In this program, the servo motor will rotate from 0 to 180 incrementally. Once it reaches position 180, it will then rotate back to 0. We use the variable position to calculate the exact position we want the servo to move to and then write the value using object write. When compiling the code, make sure you include the servo library code as this is required to access all the servo library functions and allow us to control the servo.

Taking a look at the code, there is a new function that we have not yet come across, which is an *object*.

```
Servo myservo;
```

This line of code creates a servo object called myservo; this allows you to create multiple objects and use them independently from one another.

Summary

In this chapter, we have looked at some basic outputs using various functions such as `digitalWrite` to control LEDs, LCDs, and servo motors. In the next chapter, we will look at the different types of inputs that we can use to control these outputs as well as read various sensors such as light and temperature.

5

Inputs

In this chapter, we will program some input devices such as switches, temperature sensors, and more. Input devices are generally used to trigger some output or event such as some of the devices in Chapter 4.

A simple switch can easily be used to turn something on or off, such as a light-emitting diode (LED). A sensor used as an input device can monitor something like temperature or a certain type of gas, and when triggered, it can set a series of events in motion or just be collected as data and interpreted into something visual. This chapter will also look at some of the basic programming used in reading and detecting inputs, which you can use for your own projects in the future. We will look at both digital and analog inputs, where a digital system will return a value of either 1 or 0 and an analog input will return a value from 0 to 1023 using an analog-to-digital conversion. In this chapter you will accomplish the following:

- Understand how to read digital and analog inputs.
- Experiment with using different types of input devices.
- Learn more about coding with the Intel Galileo.

Digital Inputs

When we used `digitalWrite()` in Chapter 4, there were two different states that the LED could be: either HIGH or LOW. When it comes to using digital inputs, the same states apply to the switch—it is either HIGH or LOW. When you read the state of a digital input using `digitalRead()`, it will be connected to either 3 V or 5 V, which indicates a HIGH state, or to ground, which indicates a LOW state.

Experiment 6: Reading a Switch

By connecting a simple push-button switch to the Intel Galileo, you can easily change the state of a digital input pin by pressing it.

A simple push-button switch, as shown in Figure 5.1, is great for experimenting with digital inputs and is also an inexpensive component—something to consider when building your own projects. This switch is a common component you will become familiar with, as it is included in almost every electronic starter kit. When looking at the Grove switch in Figure 5.1, the top two pins are connected, as are the bottom two pins. When the push-button switch is pressed down, both sets of pins are connected in the circuit and it becomes complete.

We are going to use a small Grove push-button switch to make or break a connection between 5 V and one of the digital pins on the Intel Galileo, which you will configure in your sketch when uploading to the Galileo. If you just connect the switch to the Galileo board, this creates a problem when the switch is not closed because the input pin is not connected to anything. This is referred to as floating and could easily give you false readings on the state of the switch. When creating a switch circuit, you need your readings to more precise, which can be accomplished by using a pull-down resistor. If you take a closer look at Figure 5.1, you may see a small surface mount resistor that is used as a pull-down resistor.

Figure 5.1 *A Grove push-button switch.*

Schematic Reference	Description	Appendix
M1	Intel Galileo	M1
	Grove base shield	H3
S1/R1	Grove button	S2

Table 5.1 *Components and Hardware*

Now think about what happens when the button is not pressed with the pull-down resistor in the circuit. The input pin will be connected through a 10K resistor to ground. Although the resistor will restrict the flow of current, there is still enough to ensure that the input pin will read a LOW logic value. 10K is a common pull-down resistor value. The value of the resistor must be low enough to make it resistant to any electrical interference, but at the same time it must be high enough so that it prevents excessive current drain when the switch is closed. When the button is pressed, the input pin is directly connected to 5 V through the button. In this circuit the current has two options:

- It can flow through a zero-resistance path to 5 V.
- It can flow through a high-resistance path to ground.

The Grove switch module already has a resistor built into the button circuit in the form of a surface mount resistor. Table 5.1 shows the components that you will need to use in this example.

Insert the Grove connector into one of the digital pins on the shield. For this example, we will also use one of the built-in LEDs on the Intel Galileo, so that when you push the switch the LED will come on or vice versa; the built-in LED on the Intel board is connected to digital pin 13. You can see the experiment in Figure 5.2 and the schematic reference in Figure 5.3.

Here is our basic digital input sketch, using the push button as the input device to light up the on-board LED, which is connected to digital pin 2 as an output.

Sketch 7—Digital Switch

```
int pushbutton = 2;
int led = 13;

void setup() {
    pinMode(pushbutton, INPUT);
    pinMode(led, OUTPUT);
}
```

Figure 5.2 *Connecting a push-button switch to the Galileo.*

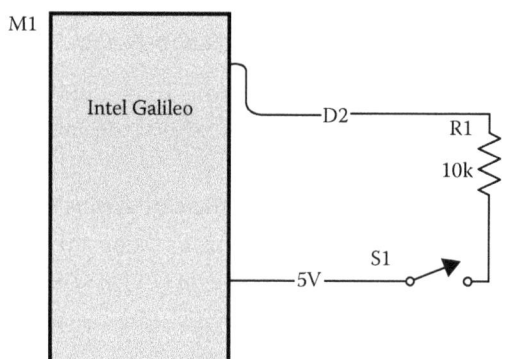

Figure 5.3 *Schematic diagram for connecting a push button.*

```
void loop() {
    int buttonstate = digitalRead(pushbutton);
    if (buttonstate == HIGH){
        digitalWrite(led, HIGH);
    }
    else {
        digitalWrite(led, LOW);
    }
    delay(100);
}
```

We can look at the code in the following sections to make it easier for us to understand what it is doing:

- Set the push button (pin 2) as an input and the LED (pin 13) as an output.
- Read the state of the push button and store it in a variable called *state.*
- If the push-button state is HIGH or connected to 5 V, turn the LED on or HIGH.
- If the push-button state is LOW, turn the LED off or LOW.
- Pause the program for 100 milliseconds to slow down the input.

When you upload this program to the Galileo and push the Grove switch, you should see the on-board LED light up. When you release the push-button switch, the LED will switch off.

digitalRead ()

The main function that we can learn from our sketch is the use of `digitalRead()`. This checks the value of the pin, which is referenced in the parameters. In this example, we refer to digital pin 2 and we call `pushbutton` to check if the push button is connected to 5 V or to ground. `digitalRead()` returns a value of either HIGH or LOW, and we store this value in a variable called *state.*

You may also note that we used `pinMode()` to set the digital pin 0 as an input in the setup function. This is required so that the Galileo board knows how to treat that pin and allows us to use the function `digitalRead()` on that particular pin.

The syntax for `digitalRead()` is
    `digitalRead(pin);`
The parameter of `digitalRead()` is
    pin; refers to the digital pin number
`digitalRead()` return HIGH or LOW

Local and Global Variables

In the previous code examples, the variables were declared outside both the `setup` and `loop` functions. Those variables are always referred to as *global variables* because they can be easily accessed and changed from either the `setup` or `loop` functions. In our example code, we declared a new variable within the `loop` function:

```
int state = digitalRead(pushbutton);
```

When you call a variable within a block of code, it can only be accessed within that block. This is known as a *local variable* because it sits within a function. When the program is running on the Galileo board and it finishes executing a particular block of code, any local variables within that block are automatically freed up in memory so they can be used for other variables in the next block of code.

Debouncing

When you press a push button, you might expect to get a single change from 1 to 0. In reality, that does not always occur. Sometimes when the contacts in the push button come together, they bounce when you release the button and create static signals. A single button press has now become two or more presses, depending on the push-button switch. All of this happens in a split second—the total amount of time the button registers its press is less than 200 milliseconds. Most new tactile-type switches may not bounce at all; however, a very old switch may have a lot of bounce. Sometimes the bouncing does not have any effect on the outcome of our sketch. For example, in our previous sketch we detected a push-button press, which then turned on an LED. Debouncing makes no difference to the outcome because when we release the switch, it will stabilize itself and the LED will switch off; this may take only a few milliseconds, so we do not notice the debouncing effect.

One situation in which debouncing may cause your outcome to be different from what is expected is using a push-button switch to turn an LED on or off every time the switch is pressed. When you press the button, the LED comes on and stays on; when you press the button again, the LED turns off and stays off. If you had a button that bounced every time you pressed it, then the LED would be on or off based on whether you had an odd or even number of bounces. Using the same circuit as our previous sketch, try flashing the following program to your Galileo:

Sketch 8—Debouncing A

```
int ledpin = 13;
int ledValue = LOW;
```

```
void setup() {
pinMode (2, INPUT);
pinMode (ledpin, OUTPUT);
}

void loop() {
    if (digitalRead(2) == HIGH) {
        ledValue = ! ledValue;
        digitalWrite(ledpin, ledValue);
    }
}
```

Press the push button a few times and see what happens. You may notice that when you push the button, the LED may not turn on or off as it should, but when you push it a few more times, it does turn on or off. As previously explained, this is a perfect example of how debouncing occurs. Try loading the following sketch with the addition of the highlighted code in bold:

Sketch 9—Debouncing B

```
int ledpin = 13;
int ledValue = LOW;

void setup() {
pinMode (2, INPUT);
pinMode (ledpin, OUTPUT);
}

void loop() {
    if (digitalRead(2) == HIGH) {
        ledValue = ! ledValue;
        digitalWrite(ledpin, ledValue);
        delay (200);
    }
}
```

Press the push button again a few times—notice anything different? This time it works fine after inserting a short delay into the program. This is because as soon as the program registers the first push of the switch, it delays the program before it checks again just in case there is another bounce on the switch.

Sometimes when writing your code, you may need to reverse a value from HIGH to LOW. You can do this with Boolean logic using the ! or not operator:

```
ledValue = ! ledValue;
```

In your program you used this to reverse the value of the LED. You set the LED value as a global variable at the start, which was LOW, so the equation is "*ledValue* is equal to not LOW," which becomes HIGH. So when we use `digitalWrite`, your value used to light up the LED is HIGH.

Analog Inputs

As you have learned with digital inputs, the information you receive from a digital device is either HIGH or LOW, ON or OFF. In contrast, there are a number of devices that can have a range of values, such as dials, sliders, temperature sensors, and many more. Analog inputs on the Galileo board give us a value ranging from 0 to 4095 up to 5 V. Just like other Arduino boards, the Intel Galileo has six 12-bit analog pins that you can use to read analog devices (see Figure 5.4).

The analog pins on the Galileo will accept any voltage between 0 and 5 V. As previously mentioned, because we are using a digital system, these values must be converted into digital using an analog-to-digital converter (ADC), which is featured on the Galileo board.

Luckily for us, we do not have to understand how the ADC works in too much detail because the software we will write will take care of most of it for us.

Figure 5.4 *Intel Galileo board analog pins.*

Experiment 7: Reading a Potentiometer

The easiest analog sensor that we can read is a simple potentiometer (POT). You can find these in most electronic kits, and they are fairly common and inexpensive; you can even salvage these from old consumer electronics such as stereos, speakers, thermostats, and more. A potentiometer is a variable-voltage divider that looks like a dial knob, as shown in Figure 5.5. They can vary in size and shape, but they all have one thing in common—they all use three pins. In our case, you connect one of the outer pins to ground and the other outer pin to 5 V. Potentiometers are also symmetrical, which means it does not matter how you connect the ground and 5 V, as long as it is not on the middle pin. The middle pin connects to the analog pin on the Galileo, as shown in Figures 5.5 and 5.6.

As you turn the potentiometer, you vary the voltage that feeds directly into the analog pin on the Galileo between 0 and 5 V. If you have a digital multimeter handy, you can confirm the value when you turn the knob—simply change your multimeter value to voltage, hook up the red probe (positive) to the middle pin on the POT, and hook up the black probe (negative) to whichever side you have the POT connected to ground.

Figure 5.5 *Grove potentiometer connected to the Galileo.*

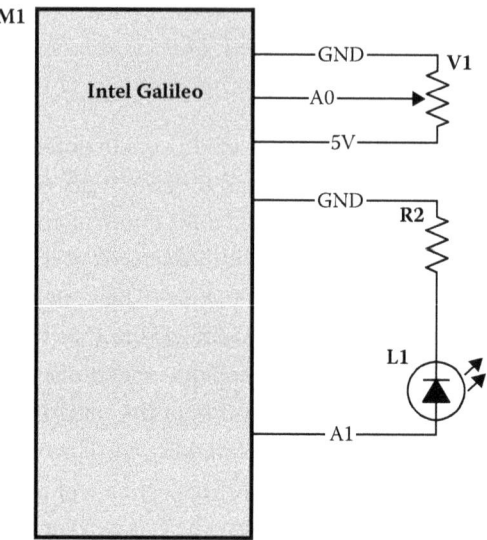

Figure 5.6 *Schematic layout diagram for reading a potentiometer.*

Schematic Reference	Description	Appendix
M1	Intel Galileo	M1
	Grove base shield	H3
V1	Grove potentiometer	R1
L1/R2	Grove LED	D1

Table 5.2 *Hardware and Components*

The hardware requirements for this experiment can be found in Table 5.2.

Here is the basic example sketch for reading analog input and using the value to create some output.

Sketch 10—Reading a Potentiometer

```
const int POT = A0;
const int led = A1;
int val = 0;

void setup() {
pinMode(led, OUTPUT);
pinMode(POT, INPUT);
}
```

```
void loop() {
int val = analogRead(POT);
int ledval = map(val, 0, 4095, 0, 255);

analogWrite(led, ledval);

delay(200);
}
```

We can take a look at the code in the following stages to understand better what it is doing:

- Create an integer called val and assign it a value of 0.
- Scale the analog read value, which runs from 0 to 4095, to the analog output value, which ranges from 0 to 255.
- Write the potentiometer value to the LED.
- Pause for a few milliseconds and repeat the process.

analogRead

Just like the function digitalRead, which returned a value of either HIGH or LOW, analogRead returns a value as well. In contrast to HIGH or LOW, analogRead returns an integer value from 0 to 4095. That value represents a voltage from 0 to 5 V.

The return value can be stored in a variable for use later on in our code, or it can be used right away using an if statement such as:

```
if (analogRead(POT) > 2000 {
      analogWrite(led, 129);
}
```

In our sketch, the analogRead value is stored in a variable called val.

The syntax for analogRead() is

```
analogRead(pin);
```

The parameter is pin, the analog pin that the sensor is connected to on the Galileo.

analogRead() returns an integer value between 0 and 4095, which represents ground and 5 V.

Const

In our code we introduced a couple of new programming concepts that we can use along the way. We previously used variables to store our pin numbers in memory; however, in our firmware we will use something called a constant:

```
const int POT A0;
```

This creates a spot of memory, just like a variable does for an integer value called POT, and stores the pin value A0. This seems like a variable but it is not—when using a constant, you cannot change or edit the value that is assigned to it, as the name suggests. This can be useful when you want to store a value in the memory and you know that this value should never be changed within the code. If you do change your code, then you will receive an error when verifying your sketch.

map()

Because the returned value of `analogRead()` is a value between 0 and 4095, you may find that you need to scale this range of values down into something a bit more manageable. In our case, we want to receive the value from the potentiometer and then write that value to the LED using `analogWrite`. Because the `analogWrite` function has a range from 0 to 255, we cannot simply write the value of the POT because it would be too high. For this we can use a function called `map()` to scale down the values to something we can use. We can scale the input value using the following code:

```
int ledval = map(val, 0, 4095, 0, 255);
```

The `map()` function is responsible for scaling one set of ranges to another. The input scale is from 0 to 4095, and the output scale is from 0 to 255. If we calculated this scaling range manually, it would become too complicated and sometimes difficult to work out; it could also return recurring values of an infinite nature.

The syntax of `map()` is

```
map(input, inform, inTo, outFrom, outTo);
```

The parameters for map() are

- **input**   the input value to be scaled.
- **inform**   the first number in the input scale.

- **inTo**   the second number in the input scale.

- **outFrom**   the first number in the output scale.

- **outTo**   the second number in the output scale.

map() returns a value on the scale of outFrom to outTo.

You can easily modify the map() function by converting a value into a percentage from 0 to 100—just change the output values.

Variable Resistors

Most analog sensors work like a potentiometer, calculating the varying voltage as the resistor changes its value. These are called variable resistors that resist the flow electricity through a circuit. A good example of this is a simple light-dependent resistor or photocell like in Figure 5.7. This changes the resistance based on the amount of light that hits it. When the light increases, the resistance goes down; therefore, the voltage in the circuit goes up. Take the light away, and the resistance increases and the voltage is reduced.

In order to read sensors like these with the Galileo board, you will need to create a voltage divider circuit and connect it to the analog input pins on the Galileo.

Figure 5.7   *A typical photocell, which acts as a variable resistor.*

Experiment 8: Voltage Divider Circuits

When you are working with different sensors that offer a variable resistor feature, you need to create something called a voltage divider circuit. A voltage divider circuit converts the variable resistance into a variable voltage, so we can read this value from the input pins on the Photon. In the schematic diagram in Figure 5.8, you can see a simple voltage divider circuit.

Figure 5.8 shows two resistors set up in series to one another between the input voltage and ground. You can also see one wire coming from between both resistors; this is the voltage output, which is the value that we read from the input on our photon board. If we first consider a fixed voltage divider, we can understand the concept of how a voltage divider works. The mathematical calculation for working out the values of a voltage divider is as follows:

Vout = Vin[R2/(R1 + R2)]

In our case, the voltage input from the Galileo would be 5 V and the voltage output would be connected to one of the analog input pins on the Galileo board.

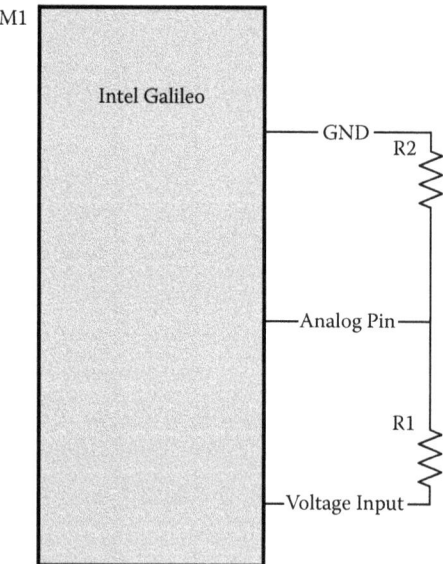

Figure 5.8   *Voltage divider circuit.*

Schematic Reference	Description	Appendix
M1	Intel Galileo	M1
	Grove base shield	H3
R3	Grove photocell	R2
L1/2/3	Grove RGB LED	D3

Table 5.3   *Components and Equipment*

If we use the resistor values for R1 and R2 so that they are matched (both 10K in this example), the 5 V is divided by 2 to make an output voltage of 2.5 V according to the equation. Let us look at this in a bit more detail by adding our values to the following equation:

$$Vout = 5\ V/[10K(10K + 10K)] = 5\ V \times 0.5 = 2.5\ V$$

Now what happens when we replace one of the resistors with a variable resistor such as a photocell? In this case we will replace resistor R1 with a 200K photocell. Whether you choose to replace R1 or R2 and which value you choose will affect the overall scale and precision of the output readings you receive. It is always worth experimenting with different configurations of resistor values until you find something that you are comfortable using and you are certain that the results are adequate.

For this example, we are going to wire up a Grove photocell and use it to determine the color of an RGB (red, green, blue) LED. Connect the circuit to your Galileo as shown in Figure 5.9 and use the contents in Table 5.3.

Sketch 11—Light Sensor RGB Indicator

```
int red = D0;
int green = D2;
int blue = D1;

void setup() {
pinMode(red, OUTPUT);
pinMode(green, OUTPUT);
pinMode(blue, OUTPUT);

}

void loop() {
    int value = analogRead(A0);
```

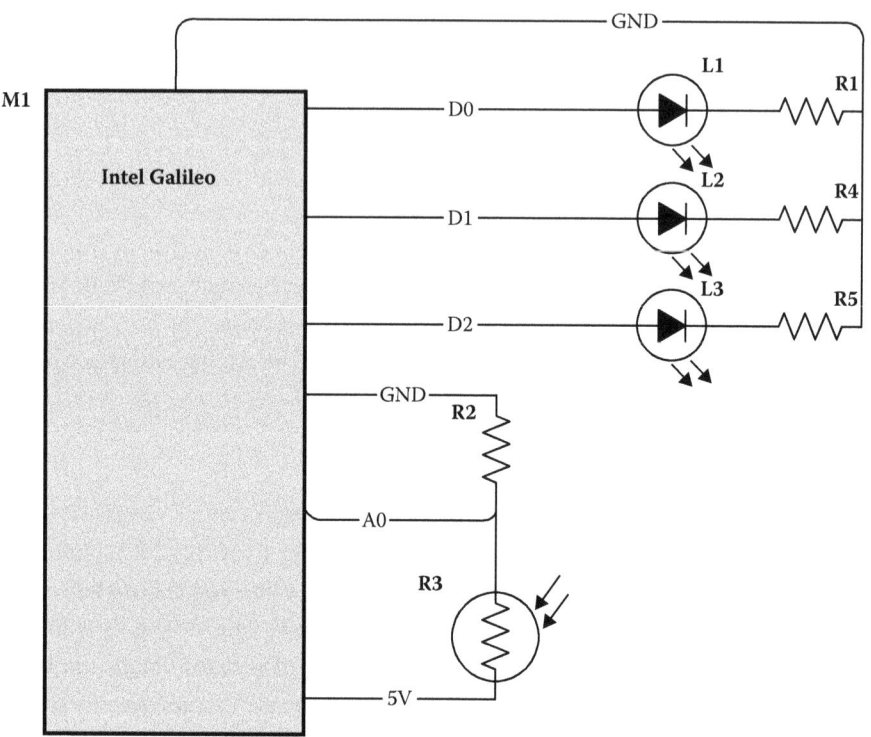

Figure 5.9 *Experiment 8 schematic reference.*

```
int percentage = map(value, 0, 4095, 0, 100);
if (percentage < 33) {
    digitalWrite(green, HIGH);
    digitalWrite(red, LOW);
    digitalWrite(blue, LOW);
}
else if (percentage > 33 & < 66) {
    digitalWrite(green, LOW);
    digitalWrite(red, LOW);
    digitalWrite(blue, HIGH);
}
else if (percentage > 66) {
    digitalWrite(green, LOW);
    digitalWrite(red, HIGH);
    digitalWrite(blue, LOW);
}
}
```

When you run the program on the Galileo, the LED should light up a particular color depending on the light levels taken from the photocell. Try covering the photocell with your hand and see if the color changes again; then try shining something bright at the photocell, and it should change the LED color. Because we are using an RGB LED, we can use a number of different combinations of colors, so we are not necessarily forced to use three different colors—we can easily use up to six with the function `digitalWrite`. However, if we use an analog output, then the number of colors and shades we can create is limitless.

We can look at the code in a bit more detail by breaking it up into the following sections. When programming code, it is always easier to break your code up into sections, as it makes it easier to understand what is going on and easier to debug your code when there are issues.

- Declare the digital pins on the RGB LED. Label these with the color that each pin will represent.

- Tell the Galileo board that the RGB LED pins are digital output pins by using the function `digitalWrite`.

- Read the input value from the photocell and store this as an integer in the variable called "value."

- Use the function `map()` to convert the input value to a percentage.

- Calculate the value using if statements to determine which color on the RGB LED to switch on and off.

- Add a short delay to the end of the code.

Within our sketch we use *if   else* functions, which allows us to check another condition if the first condition is false. The syntax looks like this:

```
if (conditions A) {
        execute the code here if A is true
}
else if (condition B) {
        execute the code here if condition A is false and condition B is true
}
else {
execute the code here if both conditions A and B are both false. This will always
be the default option
}
```

You can use as many *if else* statements as you like within the code, or you may have an else statement at the end of the chain for the code that should be executed if all conditions return false.

Experiment 9: Reading Temperature

There are many different types of temperature sensors that we can use, but keeping within the theme of using variable resistors we will use a thermistor. A thermistor changes its resistance based on temperature; therefore, we can work out the temperature from the input voltage using one of the analog pins on the Galileo board.

To measure the temperature, we need to measure the resistance. However, a microcontroller does not have a resistance meter built in. Instead, it only has a voltage reader known as an analog-digital converter. So we must convert the resistance into a voltage by adding another resistor and connecting them in series. Now you just measure the voltage in the middle; as the resistance changes, the voltage changes too, according to the simple voltage-divider equation. We just need to keep one resistor fixed.

In this experiment, we are going to use the Grove temperature sensor, which has a built-in thermistor to read the temperature value. You can refer to the circuit diagram in Figure 5.10 and the hardware required in Table 5.4.

Sketch 12—Temperature Sensor

```
const int pinTemp = A0;

const int B = 3975;

void setup()
{
    Serial.begin(9600);
}

void loop()
{
    int val = analogRead(pinTemp);

    float resistance = (float)(1023-val)*10000/val;
```

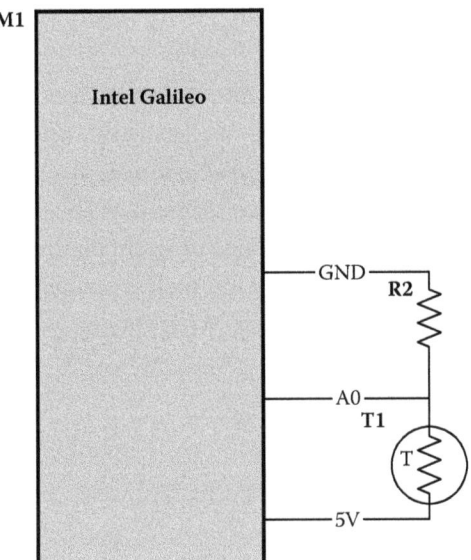

Figure 5.10 *Experiment 9 schematic reference.*

Schematic Reference	Description	Appendix
M1	Intel Galileo	M1
	Grove base shield	H3
T1	Grove temperature sensor	R3

Table 5.4 *Components and Hardware*

```
float temperature = 1/(log(resistance/10000)/
B+1/298.15)-273.15;

Serial.println(temperature);

delay(1000);
}
```

The sketch in this example requires some computational math to determine the value of the temperature in degrees Celsius. Before you start the conversion, you must know what the conversion value of your thermistor is, also known as b parameter equation value. You can find this value on the datasheet for every thermistor; in our case this value is 3975.

Summary

In this chapter, we have looked at reading input devices from both analog and digital pins on the Intel Galileo board. We have been able to read from sensors such as temperature and light, using some basic math to convert those values to a readable format. You should be quite confident now about connecting your own devices to the Galileo board and be able to program them as either an input or an output device. In the next chapter, we will look at connecting your Intel Galileo board to the network using Ethernet and Wi-Fi connections.

6

Networking: Getting Connected

Getting connected to the Internet is an important feature of most modern hardware boards, Intel Galileo included. The Galileo board does not only have built-in Ethernet network connectivity, but also features extensive support using the Arduino Ethernet and Wi-Fi libraries; furthermore, if you decide to use the full power of the Linux OS then you can open up a whole lot more outside of your Arduino sketches.

In this chapter we will look at the following topics:

* Connecting to Ethernet
* Connecting the Wi-Fi module
* Basic networking commands in Linux

Connecting to the Ethernet

By far the most convenient method of connecting to the Internet is using the built-in Ethernet port on the Galileo board. Simply follow these steps:

1. Connect your Galileo board via an Ethernet cable to your network router. You can plug the Ethernet cable directly into an available port on your router or plug the cable into an active port (Figure 6.1).

2. Connect your Intel Galileo to the power adaptor.

3. Connect the Galileo board to your computer through the USB client port on the board.

73

Figure 6.1 *Intel Galileo Ethernet port.*

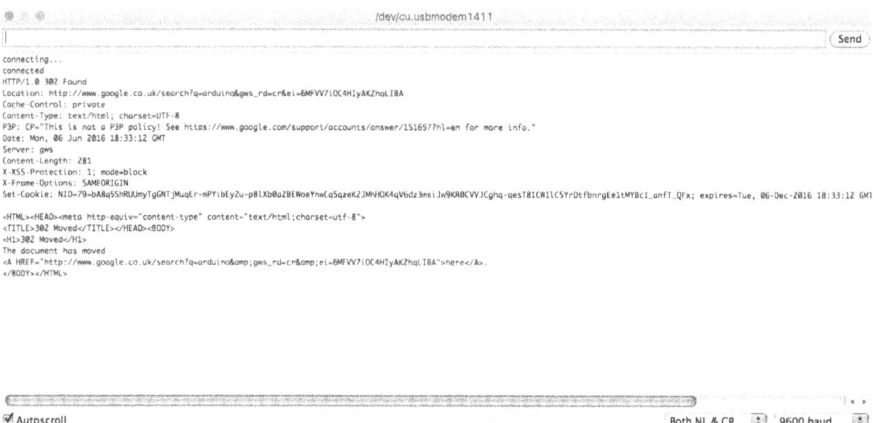

Figure 6.2 *Serial monitor output.*

4. Open up the Arduino IDE and select File > Examples > Ethernet > Web Client.

5. Click Upload.

6. Open the serial monitor to see the output from the Ethernet adaptor (Figure 6.2).

If the Intel Galileo board was successful in connecting to a server, you will see some text in the serial monitor. The Arduino Ethernet example you uploaded to the Intel Galileo board programs the board to do a search using Google for the term "Arduino." As the HTML response from the search term is received from Google, it starts to output the result to the serial monitor, which you can see.

One advantage of using the Ethernet on the Intel Galileo board is that it eliminates the requirement of purchasing an additional shield, while also supporting up to two Ethernet interfaces using the Quark SoC. In terms of using the Ethernet API, the Intel Galileo compiles with the Arduino reference using the available classes that can be implemented using the API: *Ethernet, EthernetClient, EthernetUDP, IPAddress*, and Server.

The Intel Galileo features a performance increase when using the API because the Quark SoC supports Ethernet interfaces and runs in real-time operation. When using the *Ethernet* class, the interface is configured directly on the Linux OS. Normally when running sketches, the Ethernet class checks first to see if the Ethernet has been configured; if it is not, the DHCP process is restarted, and this takes time. In the case of the Intel Galileo, because the Ethernet is already configured in the Linux OS, the Ethernet class becomes redundant; this allows more flexibility within the sketches.

Dynamic and Static IP Address

This section will show you how to configure your Ethernet network settings for both DCHP and static IP configurations. The Intel Galileo board can be configured within the Linux OS and also using the Ethernet class within your sketches. The Ethernet can be configured manually with the Linux console using the interface eth0.

Before we get started, it is worthwhile knowing how your router box is configured in order to understand the range of IP addresses used, both static and dynamic. You should be able to access your router by typing in the IP address (which can be found on the reverse side to the box) in the address box of your Internet browser; in my case it's 192.168.1.1. A password and username is usually also required to gain admin privileges and to make any necessary changes. From your router settings, you should be able to navigate through the Ethernet settings and check the IP address range for DCHP or static IP address.

Dynamic IP

Most routers these days make things easy to connect to Ethernet using a dynamic IP address system. By default, the Intel Galileo software supports dynamic IPs obtained by the Dynamic Host Configuration Protocol (DCHP). You can take a look at the configuration of the Ethernet adaptor in the /etc/network/interfaces file in the following line of code:

```
#Wired or Wireless interfaces
auto eth0
iface eth0 inet dchp
iface eth1 inet dchp
```

The eth1 configuration is irrelevant in this case because the Galileo will only ever support one Ethernet adaptor. With this DHCP configuration, all that is left to do is connect your Ethernet cable from the Intel Galileo to your router. You can enable and disable the Ethernet adaptor on the Intel Galileo by tying `ifdown eth0` and `ifup eth0` in the console terminal.

Once connected, you can run a simple test from the command line that will test your Ethernet connection to the Internet. A simple ping to Google will test the connection; if successful, you will see a response from the Google servers.

```
ping www.google.com
```

Static IP

A static IP address on the Intel Galileo can be useful in two scenarios. First, if you connect to your Galileo board through SSH then you will login using the IP address; however, if the address keeps changing every time you boot up the Galileo then you will be constantly checking your router configuration to see what the IP address is for you to connect. Second, if your router is actually too far away for a cable to reach, then you might want to connect it to your local computer.

You can configure a static IP address to the Intel Galileo using a direct connection to your computer to either transfer files or even share your computer's Internet connection. If you decide to share your computer's Internet connection with the Intel Galileo board, then you need to identify your Ethernet adapter connected to the Galileo board; this configuration must be done on your computer. For example, if your Galileo board is connected to your computer's Ethernet and your Internet is connected to your wireless router, then you need to tell your computer to share the wireless Internet with the Ethernet adaptor.

Windows 7 Configuration

The configuration on a Windows 7 computer is actually quite a simple process; you have to create a local connection and configure the IPv4 settings using a static IP address.

1. Access your network settings through Control Panel > Network and Internet > Network and Sharing Center.

2. Click on change adaptor settings and right-click on Local Area Connection (Figure 6.3).

3. Select IPv4 and click the Properties button. The IPv4 configurations settings will be shown as in Figure 6.4. In this field, add a valid static IP address, subnet mask, and gateway, which is usually your router's IP address.

4. The subnet mask must be the same as your router; otherwise, the connection will not be on the same network number. You can find the subnet mask from the command shell by typing ipconfig.

5. Most routers will have a default subnet mask of 255.255.255.0.

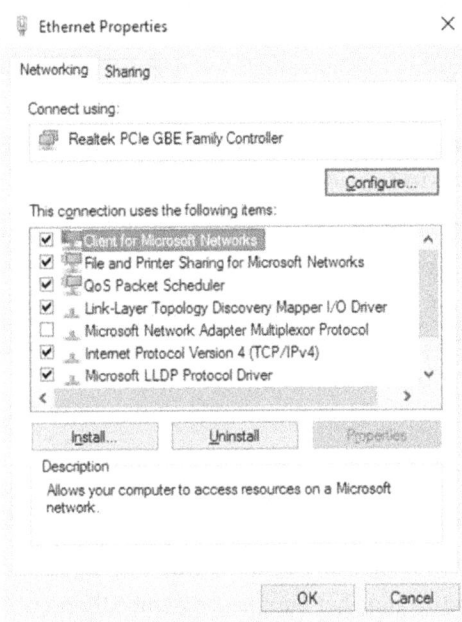

Figure 6.3   *Local area connection properties.*

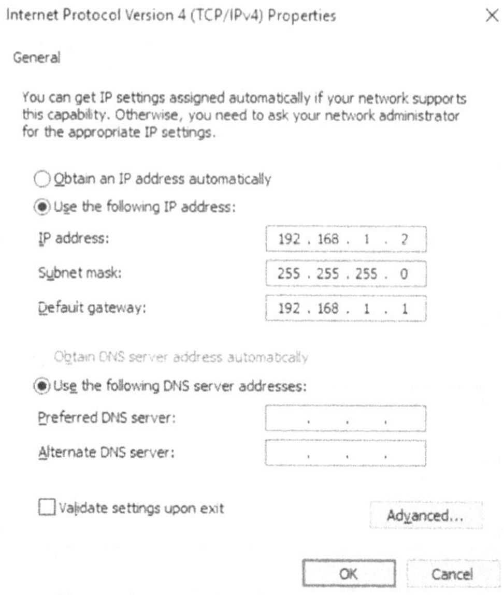

Internet Protocol Version 4 (TCP/IPv4) Properties   ✕

General

You can get IP settings assigned automatically if your network supports
this capability. Otherwise, you need to ask your network administrator
for the appropriate IP settings.

○ Obtain an IP address automatically
◉ Use the following IP address:

IP address: 192 . 168 . 1 . 2

Subnet mask: 255 . 255 . 255 . 0

Default gateway: 192 . 168 . 1 . 1

○ Obtain DNS server address automatically
◉ Use the following DNS server addresses:

Preferred DNS server: .   .

Alternate DNS server: .   .

☐ Validate settings upon exit   Advanced...

OK   Cancel

Figure 6.4   *IPv4 properties.*

6. To share your Internet connection from your computer, navigate to your
   network adaptors by choosing Control Panel > Network and Internet >
   Network and Sharing Center and click Change Adaptor Settings. Now
   right-click in the adaptor of your computer that provides the Internet,
   such as your wireless network adaptor. Click the Sharing tab and check
   the option to share the Internet in the list box, Figure 6.5.

At some point Windows might show you a warning message stating some
other static IP address will be associated with your connection. This is because
Windows does reserve some IP addresses for sharing. Just accept the new IP and
connect the Ethernet cable to the Galileo board.

MacOSX

The first thing that you will need to do before setting up a static IP address is
determine if your Mac has an Ethernet adaptor; if it does not, then you will
need to consider purchasing one. Modern Macs will require an adaptor that has a
Thunderbolt connector and that supports gigabit Ethernet. There are of course
other options, such as USB adaptors, but in terms of performance the Thunderbolt adaptor is the best option at around $30.

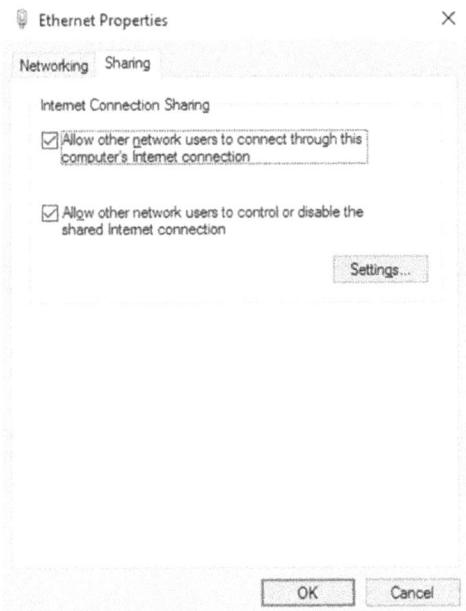

Figure 6.5 *Sharing your Internet.*

1. Disconnect any Ethernet cable.

2. Click on System Preferences and then on Network.

3. Change the IPv4 drop-down box to manually and then add the network details such as IP address, subnet mask, and gateway, which is a named router in this case.

4. Click the Advanced button and select the DNS tab to configure the DNS server as shown in Figure 6.6.

5. Connect the Intel Galileo to your Mac using the Ethernet cable; once connected you should see the status change to connect with a green dot.

6. At this point you may want to share your Internet connection between adaptors. Using the Apple menu, choose System Preferences and select the Internet Sharing. Select the network adaptor that will be sharing the Internet and then the adaptor that will be receiving the Internet sharing. For example, you may select your wireless adaptor the Internet is on and then the Ethernet adaptor to which the Galileo board is connected.

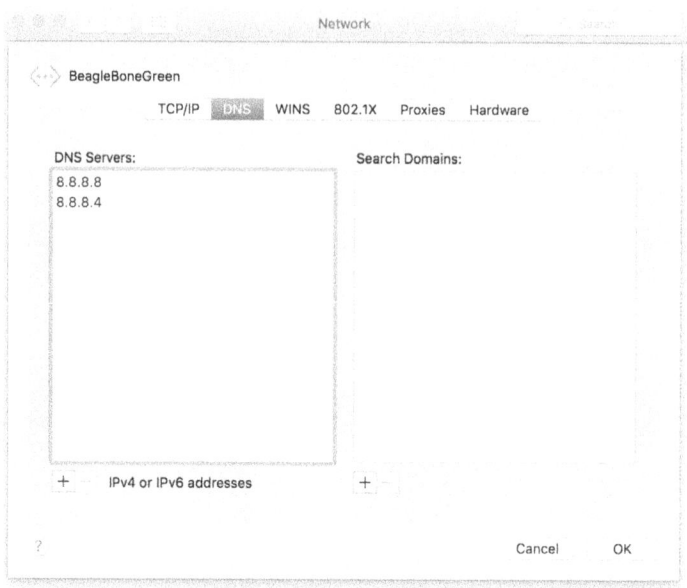

Figure 6.6 *Configure DNS settings.*

Configure Galileo Static IP

Now you should have your computer set up and connected to the Intel Galileo board through the Ethernet cable. The static IP address on the Intel Galileo can be configured manually using the command line in the console. To configure the Ethernet manually, type the following in the console:

```
Ifconfig eth0 192.168.1.27 netmask 255.255.255.0 up
```

You can test your connection by using the ping tool to send and receive commands from your computer:

```
Ping 192.168.1.22
```

Once you have the Ethernet network setup, you can add the static IP address to the network configuration file in */etc/network/interfaces*, so when the Intel Galileo board powers up, it will automatically enable the Ethernet adaptor and connect using the static IP address set. Add the following line to the file:

```
auto eth0
iface eth0 inet static
address 192.168.1.27
```

```
netmask 255.255.255.0
gateway 192.168.1.1
```

Configuring Static IP in the Arduino Sketch

The configuration of the IP address in the Arduino sketch is quite a simple process and is very similar to the command you put in the console. Add the following lines of code into the setup function of the sketch:

```
System("ifconfig eth0 192.168.1.28 netmask 255.255.255.0
ip");
System("route add default gw 192.168.1.1 netmask
255.255.255.0");
```

Connecting to Wi-Fi

If you want to use the Wi-Fi on your Intel Galileo, you will need to purchase either a mini-PCIe card with adaptor that slots in to the adaptor on the reverse side of the Intel Galileo board or a USB wireless dongle that can plug into the USB host connector on the Galileo. If you are going to use a USB dongle, make sure that the drivers are available for use with Linux; also check which mini-PCIe wireless adaptors are compatible with the Intel Galileo board.

Using a wireless adaptor is very advantageous as it allows the board to be somewhere where you physically cannot, and you can still gain access to the Galileo through a remote connection. The following section will show you how to setup your wireless adaptor and configure the connection settings to access the wireless network of your router.

Setting Up the Wi-Fi Mini-PCIe Card

In addition to sourcing the mini-PCIe card, you will also need the following items in Table 6.1.

Description	Appendix
Intel Wi-Fi mini-PCIe module	H7
Dual-band antennas	H8
Half to full height mini-PCIe card bracket	H9
4-GB micro-SD card (minimum)	H10

Table 6.1 *Hardware Required for Wi-Fi Mini-PCIe*

1. Make sure that the Intel Galileo board is powered off; remove both the USB programming cable and the power adapter.

2. Install the half to full height bracket. The Wi-Fi module should be screwed into the adaptor bracket.

3. Connect the dual-band antennas in the card connector by pressing them down until you feel a small click in the connector.

4. Insert the Wi-Fi mini-PCIe card at an angle, making sure the card's contacts are connected to the slot.

5. Download the SD card image from the Intel Galileo website: https://communities.intel.com/docs/DOC-22226

6. Transfer the downloaded image to a micro-SD card, making sure that it is FAT32 formatted. Insert the micro-SD card in the Galileo's SD card slot.

7. Connect the power supply and then the USB cable.

If you are following all the steps correctly and the mini-PCIe card was inserted, type ifconfig in the command terminal, and you should see a new adapter listed under wlan0. At this point you can create some sketches in the Arduino IDE. There are several examples in the Wi-Fi API in File > Examples > Wi-Fi. There are two examples that are of interest: one that scans the wireless networks and the other that connects to one of the wireless networks. You can refer to the Arduino reference at http://arduino.cc/en/Reference/WiFi .

You can set up the wireless interface to automatically connect the wireless network from the command terminal. You will be required to know the name of your router (SSID) and the passcode to access the Wi-Fi. You can edit the network file by typing the following:

```
Nano /etc/network/interfaces
```

The auto wlan0 adaptor should be added in here in the beginning of *iface wlan0*. Edit the files and add the following lines:

```
# Wireless interfaces
auto wlan0
iface wlan0 inet dhcp
iface wlan0 inet dchp
        wireless_mode managed
        wireless-essid your network id
```

```
wireless-key 123456789
wpa-driver wext
```

Save the new configuration and restart the connection by typing the following command:

```
/etc/init.d/networking restart
```

Getting Started with Intel XDK IoT Edition

In this section, we will look at the setup and installation of the Intel XDK IoT Edition, upload one of the examples to the Intel Galileo board, and create our own real-time dashboard in Node.js. Intel XDK IoT Edition is great for creating Web interfaces for the Galileo and supporting a variety of cloud connectivity between devices.

Requirements:

- A desktop or laptop computer running Windows, Mac OSX, or Linux.

- Your Intel Galileo board connected to your network.

- Grove Starter Kit: Intel IoT Edition (optional).

Download and Install Intel XDK IoT Edition

1. Visit the IDE download page to download a copy of the Intel XDK IoT Edition installer for your desktop or laptop computer. Make sure you select your OS platform from the dropdown list, then click the download button.

2. Run the installer by one of the following methods:

   a. Windows—Right-click the installer, then select *Run as administrator*. If a confirmation message is displayed, click *Yes* to continue.

   b. Mac—Double-click the .dmg file to extract the installer. Double-click the .pkg file to start the installation process.

   c. Linux:

      i. Open terminal.

      ii. Navigate to the folder where the installer is saved i.e. `cd ~/Desktop/`.

      iii. Extract the install files, enter the following: `tar zxvf filename`.

      iv. Navigate to the folder holding the extracted files.

      v. Run the installer; enter the following: `./install.sh`.

3. Follow the instructions in the installation wizard to install the Intel XDK IoT Edition.

Download and Install Bonjour Print Services (Windows OS)

If you are a Windows user, you may be prompted to install the Bonjour Print Services on your computer, which allows the Intel XDK IoT Edition to automatically detect any IoT devices on your local network such as the Intel Galileo board. This makes connecting to your Galileo a lot easier; alternatively, you can still enter your board's IP address and any necessary login information.

1. Go to the Bonjour Print Services for Windows page: http://support. apple.com/kb/DL999.

2. Click Download.

3. On your computer, right-click the downloaded file, BonjourPSSetup.exe, then select Run as Administrator. If you are prompted for confirmation, click Yes.

4. Follow the instructions in the installation wizard to install.

Creating a New Project

In this example, we will create a new project from one of the templates in the Intel XDK and upload it to the Intel Galileo board.

1. Run the Intel XDK IoT Edition by double clicking the icon in the Windows menu or desktop.

2. Follow the on-screen instruction to login to your Intel XDK account, or sign up for a new XDK account if you have not yet registered.

3. In the Projects tab, click Start a New Project in the bottom left once logged in. The new project page will open.

Figure 6.7 *List of project templates in Intel XDK.*

4. Click Templates in the Internet of Things Embedded Application drop-down list on the left. You should see a list of project templates on the right hand side; see Figure 6.7.

5. Select the on-board LED Blink in the templates, then click continue at the bottom of the page.

6. Select the project directory in which you wish to save your project; the default directory is Documents.

7. Type the name of your new project in the Project Name field; see Figure 6.8.

8. Click Create in the final step, and your project is created.

Connecting to the Board

1. From the IoT device drop-down list in the bottom left, select your development board if shown. Alternatively, select Add Manual Connection and type your Intel Galileo's IP address in the field.

2. If you have created a username and password to login to your board, type them in the fields on the next page; otherwise, just leave the default values in the fields.

3. Click Connect.

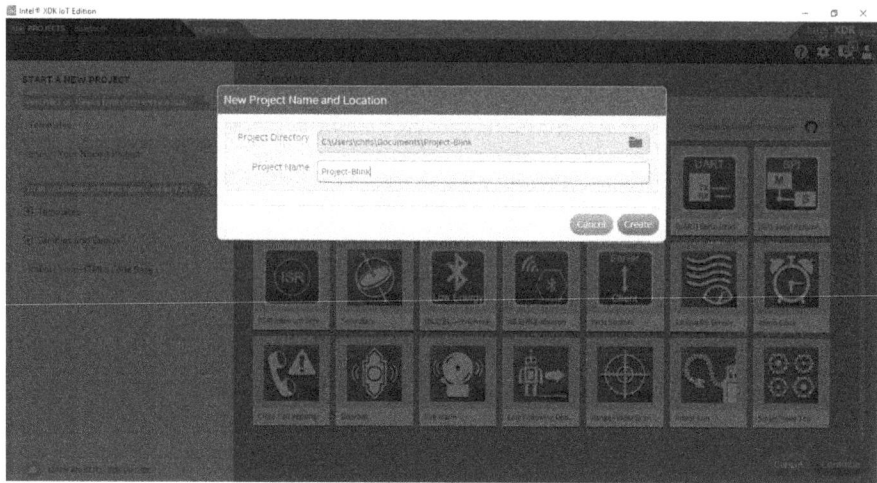

Figure 6.8 *Naming your project.*

4. If you get a message notifying you that your daemon is out of date, update it. This may disconnect you from the board as it needs to restart. Follow any instructions and reconnect when done.

5. You will also be prompted to sync the clock on-board the Galileo; just click Sync.

6. Eventually a confirmation message will appear, displaying the connection status and the IP address of the Galileo board; click dismiss and you should now be connected.

Uploading and Running the Project

1. From the top the application window, click the Develop tab. You should see the IoT toolbar at the bottom of the window as shown in Figure 6.9.

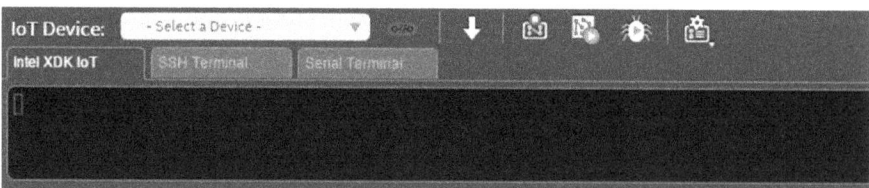

Figure 6.9 *IoT toolbar.*

2. Click the Upload icon to upload your project to the Galileo.

3. Now click the Run icon to run your code. You should now see the on-board LED blinking on and off.

4. To stop the program simply click the Stop icon.

Summary

In this chapter, we have learned how to connect the Intel Galileo to the Internet using both the Ethernet interface and the Wi-Fi module. You have learned not only how they work but also how to configure them. More importantly, you can now connect to your network through a variety of means such as using static and dynamic IP addressing. You should now be familiar with installing the Intel IoT XDK Edition on your computer after uploading one of the example programs through your local network to the Galileo board.

In the next chapter, we will create our own application to monitor and upload sensor data to the Web using a cloud platform called dweet.io. We will also use another cloud platform to take the raw data and provide a more visual to display that you can share with others.

7

Internet-Connected Hardware

If you followed the previous chapter, then you should now have your Intel Galileo board connected to your local network and also the Internet. In this chapter, we will be looking at how we can get connected hardware from the Intel Galileo to the cloud using a cloud platform called dweet.io.

Dweet.io is a simple publishing and subscribing platform for sensors and other data. The methodology has been adopted from the popular social network platform called Twitter (as you can guess from the name dweet.io). Dweet.io works in a very similar way to Twitter by posting the messages that you publish to its platform on the Web for everyone to see and use. I know there are a lot of other platforms on the market such as Amazon Web Services, IntialState, and Intel Analytics, but dweet.io is just a simple platform that works without any API reference or subscription requirement.

In this chapter, we are going to use it to post data from a number of Grove sensors such as light and temperature sensors.

Up to now we have been doing most of our programming using the Arduino IDE, but to utilize the full power and potential of the Intel Galileo board we are going to use the Intel IoT XDK Edition to create some projects using Node.js, which is a JavaScript-based runtime designed for network applications. If you remember, in the previous chapter we did touch base on this when we uploaded the Blink LED example. The Blink example was using JavaScript, which was then uploaded and run on the Galileo board remotely.

Let us take a quick look back at the code to see what is happening:

```
var mraa = require('mraa');

var myOnboardLed = new mraa.Gpio(13);
myOnboardLed.dir(mraa.DIR_OUT);
var ledState = true;

periodicActivity();

function periodicActivity()
{
  myOnboardLed.write(ledState?1:0);
  ledState = !ledState;
  setTimeout(periodicActivity,1000);
}
```

The first line of code requires us to import a library called mraa. This library is a low-level skeleton library, which allows Linux platforms such as the Intel Galileo to interface with its GPIO pins. Therefore, this library is required in the code as we are accessing the on-board LED, which as we know is on digital pin 13.

```
var mraa = require('mraa');
```

The second line of the code sets the LED pin number with a variable called *myOnboardLed*. This is similar to naming your pins in the Arduino IDE and then only referencing the name of the pin rather than the pin number itself.

```
var myOnboardLed = new mraa.Gpio(13);
```

Just like any Arduino sketch, you need to set the direction of the pin, whether this is an input or an output pin. Then the next line of code sets the GPIO pin as an output device.

```
myOnboardLed.dir(mraa.DIR_OUT);
```

As the example is blinking an LED on and off, we need to get the state the LED is in before we decide whether to turn it on or off. For this we need to create a variable which has a Boolean state.

```
var ledState = true;
```

The next line calls a function called *PeriodicActivity*. This works the same as calling any function within an Arduino sketch.

```
periodicActivity();
```

The next block of code is the function itself, which will blink the LED on or off depending on its state. If the LED state is True, then we write the LED value to HIGH; otherwise, we write the value to LOW.

```
myOnboardLed.write(ledState?1:0);
```

After you switch the LED on or off, you need to reverse the state of the variable with the following line.

```
ledState = !ledState;
```

And, finally, we set a timeout period in milliseconds before the function is then called again.

```
setTimeout(periodicActivity,1000);
```

Now you can see just how simple it is to create JavaScript code; with just a few simple lines, you can control the GPIO pins on your Intel Galileo board. You can also see the similarities between the JavaScript code and your Arduino sketches, and there are many. So, it may seem a bit daunting at first if you are unfamiliar with the programming language, but you should actually pick it up quite rapidly as we progress through the book.

Experiment 10: Analyzing Sensor Data on the Cloud

In this experiment, we are going to be using a couple of Grove sensors connected to the Intel Galileo board and then push the values we read from them to the cloud platform dweet.io. I will break down the code into sections for each sensor so that you can understand what each block of code is doing and the syntax that goes with it. In this experiment, we will be using the Grove temperature, light, and also air quality sensors, which all use the analog pin on the Grove shield. You can of course connect your own hardware using a breadboard if you wish to do so. The hardware requirements for this experiment can be found in Table 7.1.

To set up the Intel Galileo board and connect all the hardware, follow these steps:

1. Connect the Grove base shield to the Intel Galileo by inserting it into the Arduino header pins.

Description	Appendix
Intel Galileo board	M1
SD card with latest IoT image	H10
Grove base shield	H3
Grove temperature sensor	R3
Grove light sensor	R2
Grove air quality sensor	R4

Table 7.1   *Components and Hardware for Experiment 10*

2. Connect the Grove sensors to the shield's analog pins, making note of which sensor is connected to which pin. For this experiment, I connected the following:

    a. Grove air quality sensor—analog pin 0

    b. Grove light sensor—analog pin 1

    c. Grove temperature sensor—analog pin 2

3. Insert the micro-SD card with the Intel IoT image on, into the micro-SD slot.

4. Power up the Intel Galileo.

Now that the Intel Galileo is set up, we can start a new project in the Intel XDK IoT on your computer. Create a new project by navigating to the Internet of Things Embedded Application templates and locating the Blank Template. Once highlighted, click continue in the bottom of the toolbar, as shown in Figure 7.1.

Figure 7.1   *Creating a new project.*

When the new project is created, you should see the main.js file in which we will input the following code.

Experiment 10 JavaScript Code:

```javascript
var mraa = require("mraa");
var request = require('request');

var C = 4275;

var air_sensor_pin = new mraa.Aio(0);
air_sensor_pin.setBit(12);
var temp_sensor_pin = new mraa.Aio(2);
temp_sensor_pin.setBit(12);
var light_sensor_pin = new mraa.Aio(1);
light_sensor_pin.setBit(12);

function measure_data() {

    // Measure light
    var a = light_sensor_pin.read();
    var light_level = a/4096*100; // Analog input is on
4096 levels
    light_level = light_level.toPrecision(4);
    console.log("Light level: " + light_level + " %");

    // Measure temperature
    var b = temp_sensor_pin.read();

    var resistance = (4096 - b) * 10000 / b;

    var temperature = 1 / (Math.log(resistance / 10000) /
C + 1 / 298.15) - 273.15;
    temperature = temperature.toPrecision(4);
    console.log("Temperature: " + temperature + " C");

    // Measure Air Quality
    var c = air_sensor_pin.read();
    var air = c / 4096 * 100;
    air = air.toPrecision(2);
    console.log("Air Quality: " + air + " %");


    // Send request
    var device_name = 'Chris Rush'; // Change with your
own name here
```

```
    var dweet_url = 'https://dweet.io/dweet/for/' +
device_name + '?temperature=' + temperature + '&light=' +
light_level + '&air=' + air;
    console.log(dweet_url);

    var options = {
      url: dweet_url,
      json: true
    };

    request(options, function (error, response, body) {
      if (error) {console.log(error);}
      console.log(body);
    });

}

measure_data();
setInterval(measure_data, 10000);
```

The first section of the code is where you set the variables and pin numbers for each one of the sensors. The first variable sets the inclusion of using the "mraa" library, which allows us to access the GPIO pins on the Intel Galileo board. The second library "require()" allows us to load local modules and send them across HTTP; we use this to send the sensor data to the cloud.

You will probably notice another line below where we set the pin number for the sensors. This line is required to set the bit resolution of analog pins on the Intel Galileo board, which gives us a maximum value of 4095 when reading the analog pins. We also set another variable value that we will use when calculating the temperature further on.

```
var mraa = require("mraa");
var request = require('request');

var C = 4275;

var air_sensor_pin = new mraa.Aio(0);
air_sensor_pin.setBit(12);
var temp_sensor_pin = new mraa.Aio(2);
temp_sensor_pin.setBit(12);
var light_sensor_pin = new mraa.Aio(1);
light_sensor_pin.setBit(12);
```

Next we create a function that we call at the end of our program. This function breaks down a number of things into four sections: measure light levels, measure temperature, measure air quality, and, finally, send the data. The first sensor we will read is the light sensor, which is an analog sensor. First we read the raw value of the pin using the *pin.read()* function. Once the value is returned into the variable, we can then calculate this as a percentage using some basic math. The raw value is divided by the maximum value of 4096 and then multiplied by 100 as the percentage.

Because the value returned is a float value, it has a number of decimal places for high precision; however, we only want to display four digits with two decimal places using the function *to.Precision(4)*.

```
var a = light_sensor_pin.read();
    var light_level = a/4096*100;
    light_level = light_level.toPrecision(4);
```

Reading the temperature sensor is very similar to the way we get the raw value; however, there is a more complicated mathematical equation to convert the value into degrees Celsius. Before the conversion can take place, you need to know the resistance value that the NTC temperature sensor is providing.

```
var resistance = (4096 - b) * 10000 / b;
```

The next line of code is the calculation from the value read into degrees Celsius. We also use *Math.log* function here to get a logarithmic value returned.

```
var temperature = 1 / (Math.log(resistance / 10000) /
C + 1 / 298.15) - 273.15;
```

The temperature value that is returned in the variable "temperature" is then converted into a value with two decimal places or four digits total, which is an accurate temperature value to display.

The air quality sensor is the same code as the light sensor. We simply re-use the code to read the analog value of the sensor and then convert this into a percentage to determine the quality of the air.

```
var c = air_sensor_pin.read();
var air = c / 4096 * 100;
air = air.toPrecision(2);
```

The value returned is written as a whole percentage value with no decimal places.

Now that we can read all of our sensors and get the values from them, we can turn our attention to sending the values to the cloud service dweet.io. To do this, we must first make sure we have imported a Node.js API called require at the start of our program.

```
var request = require('request');
```

Request is an agent designed to make simple HTTP request calls through your program. We use this API to send our data to dweet.io. As explained previously, dweet.io is a simple cloud service that displays the data you send and allows others to view the feed. All we need to send to the cloud service is a unique name, which we can identify in the feeds and the data we wish to send. First, we set a couple of variables before we send the data. The first is the unique name that will be displayed in the dweet.io feeds; this can be absolutely anything you wish but must be relevant to you. For this example, I have used my name. The second variable is the URL to which we send the request. The URL must feature two things for it to work; take a look at the following example:

```
https://dweet.io/dweet/for/my-thing-name?hello=world
```

The first parameter comes after https://dweet.io/dweet/for, and this always must be the unique name of the feed you send to dweet.io to enable us to find it. The second parameter to follow is the value, which you are going to send to dweet. io to display along with the name of that value. Since we are sending temperature, light, and air quality, we will give these names to the names of the values sent.

```
var device_name = 'Chris Rush'; // Change with your own
name here
var dweet_url = 'https://dweet.io/dweet/for/' + device_
name + '?temperature=' + temperature + '&light=' + light_
level + '&air=' + air;
```

The next variable sets up the options to pass through the http request; this includes the URL that we created earlier with our values to send and also that sends the value as a JSON object, which is a format for readable text based on the JavaScript notation.

```
var options = {
    url: dweet_url,
```

```
    json: true
};
```

Next, we actually send the data using the "request" API, which passes all the variables and also returns a response from the server to say everything was OK.

```
request(options, function (error, response, body) {
     if (error) {console.log(error);}
     console.log(body);
});
```

Last but not least, we call the function to measure the data and send the data with the new updated values, and we set a period interval to send it every 10 seconds.

```
measure_data();
setInterval(measure_data, 10000);
```

That pretty much sums up the code for this experiment. It is a simple experiment to get you familiar with using JavaScript and using the request API to send the HTTP request to the dweet.io feed.

On your computer now, open up your Web browser and type the following in to the URL to view your feed:

```
https://dweet.io/follow/Chris%20Rush
```

Where my name is you just type the name of your feed as we set in the code; if there is a space between characters, then you need to add "%20" for each space in the URL. Here you should see the feed as shown in Figure 7.2.

Creating a Visual Dashboard

Now that we have a feed to dweet.io for our data, we can use this feed to create a visual dashboard using a platform called freeboard. Freeboard is a free cloud service that takes feeds from the likes of dweet.io and other sources and allows you to create a number of visuals to display that data.

1. Open your Web browser on your computer and navigate to www. freeboard.io.

2. Create an account, filling in your credentials such as email address and password. Creating an account is free of charge.

3. Login in to your new account and click on "My Freeboards" in the tool bar.

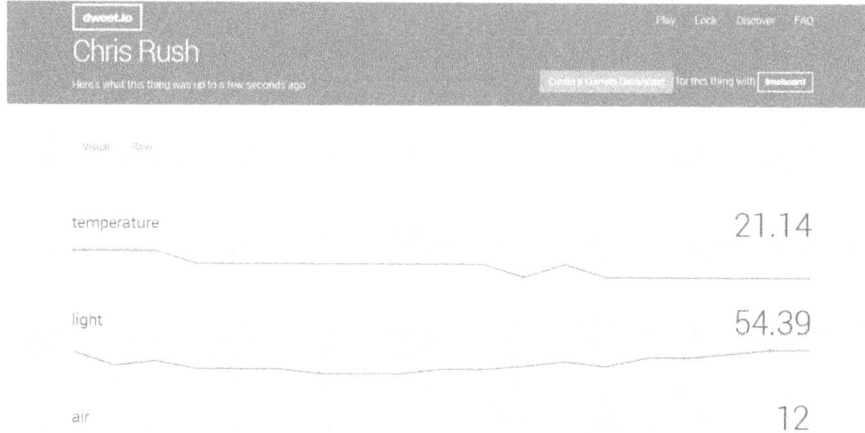

Figure 7.2 *Dweet.io feed for reading sensors.*

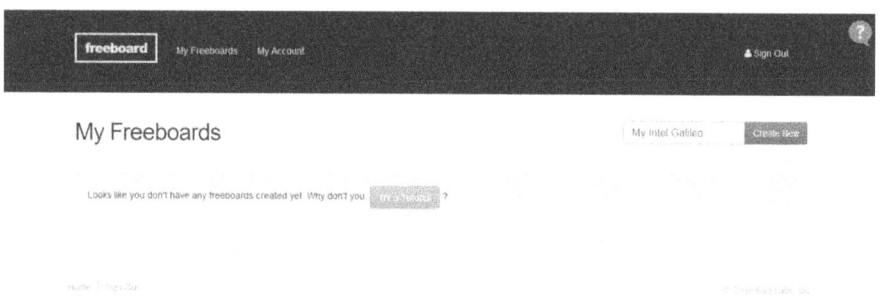

Figure 7.3 *Creating a new dashboard.*

4. Create a new dashboard and name it something like "My Intel Galileo," as shown in Figure 7.3.

5. To create a visual dashboard you must first add your data source to the dashboard by clicking "ADD" under Data Sources (see Figure 7.4).

6. Select "dweet.io" as a type of source from the drop-down box (see Figure 7.5).

7. You should be presented with some options and fields to fill in. First select a name for the data source such as "Intel Galileo"; in the text field labeled "Thing Name" input your unique name for the dweeet.io feed,

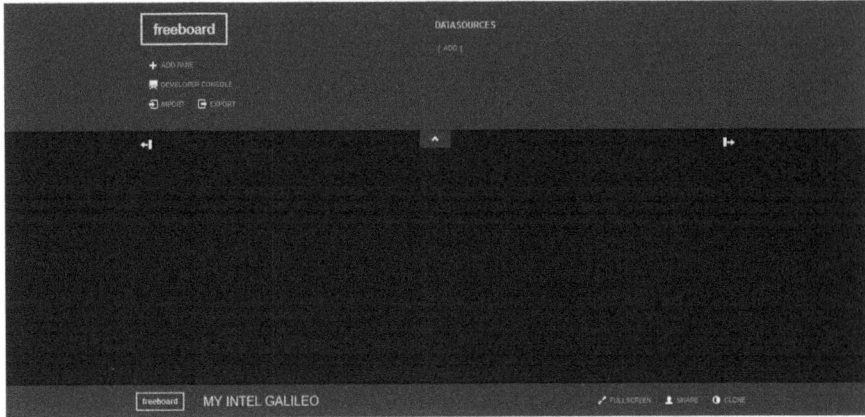

Figure 7.4 *Add a new data source.*

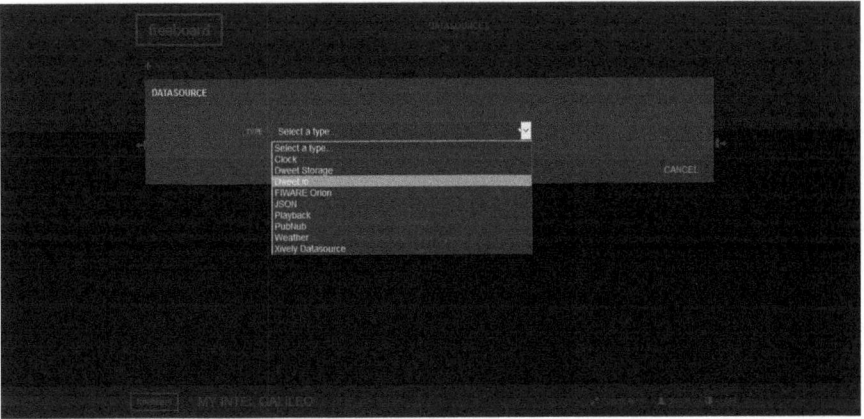

Figure 7.5 *Selecting dweet.io as a type of source.*

which in my case is my name "Chris Rush" (see Figure 7.6). You can ignore all the other fields as these are not required; go ahead and save the settings.

8. Select "Add Pane" from the toolbar to create a visual from the data feed. This will add an empty window into your dashboard.

9. Click the plus symbol from within that new pane (see Figure 7.7).

10. For the first dashboard element we are going to add a "Gauge" to show the percentage of light. Select Gauge from the drop-down list. Give

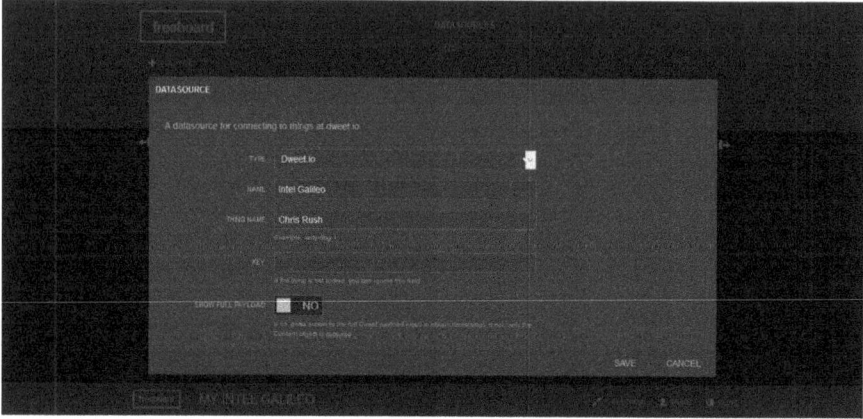

Figure 7.6   *Complete the fields for the dweet.io data source.*

Figure 7.7   *Add an empty pane to your dashboard.*

your pane the name "Light Level"; click "DataSource," click your device "Intel Galileo," and select "light," which is the name of the feed in dweet. io as set per the program code (see Figure 7.8).

11. As we are displaying the light level as a percentage, you can add the percentage symbol to the unit's field; the minimum and maximum can remain the same (see Figure 7.9).

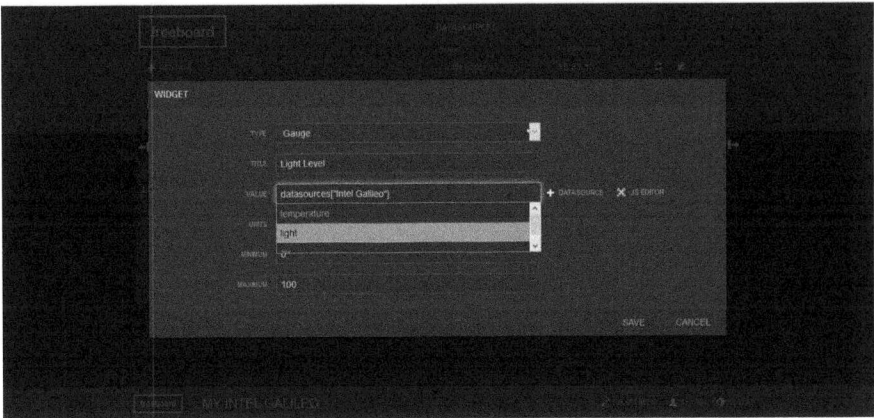

Figure 7.8. *Select the feed for the gauge.*

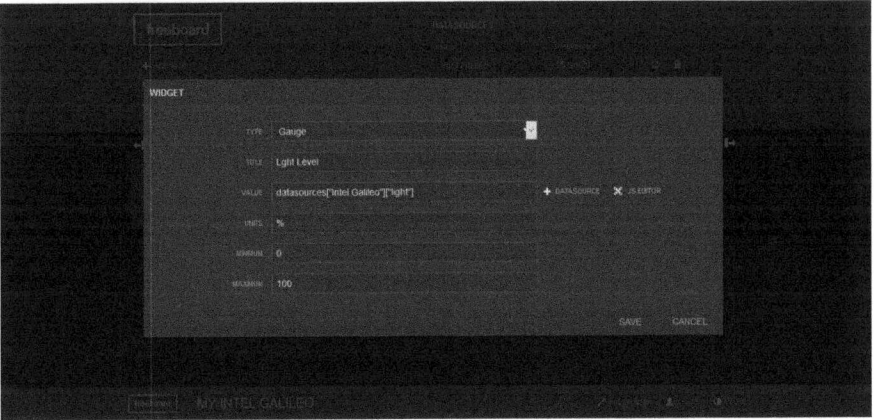

Figure 7.9 *Complete the widget fields.*

12. Click Save; you should see a Gauge created in your dashboard showing
    the light level from the sensor as shown in Figure 7.10. The value
    should update every 10 seconds according to our program.

At this point you can go ahead and create the other widgets on your own; be
sure to play around with some other ways to display your data such as plain text
or a chart. You should eventually see something like Figure 7.11 that displays all
the sensor data.

Figure 7.10 *Light level widget.*

Figure 7.11 *Freeboard dashboard.*

Experiment 11: Creating a Simple Web Server

In this experiment, we are going to create a simple Web server on our Intel Galileo board. It is relatively easy to run a Web server on the board because the Intel Galileo's processor runs a simple Linux distribution. In this experiment we will create a simple Web server using the Node.js framework along with using the Intel XDK software, which we used in the previous experiment. The combination of

Figure 7.12 *Creating a new project.*

the two makes life a lot more simple, and you will see that it is really simple to run a Web server on the Galileo.

First, you will need to install the Intel XDK software, if you have not done so already. You can refer to the beginning of this chapter to learn how to install and connect it to your Intel Galileo board. Before we can write the code for our server, we need to create a new project in the Intel XDK Software. This will allow us to write the code inside the XDK and upload it to the Intel Galileo board. For this experiment, we will write our code from scratch, even though there is a template for this project we could have adapted; but to learn and understand how it works better, I will walk you through the process step-by-step.

Create a new blank template as shown in Figure 7.12 and give it a name; for this experiment I have called the project "WebServer."

We are now going to create a simple "Hello World" Web server. When we navigate to the Web URL, the client then sends a request to the server to send it some data to display. In this example, we are going to send the term "Hello World." If we receive this response, we know that everything is set up and is working correctly. Load the following code into your main.js file:

```
var http = require('http');

var server = http.createServer(function (request,
response) {
```

```
  response.writeHead(200, {"Content-Type": "text/
plain"});
  response.end("Hello World\n");
});

server.listen(3000);

console.log("Server started!");
```

Click the Upload button on the toolbar to upload the code to the board. Then, click on the Run button, which is near the Upload button on the same toolbar. This will then start the Web server we have just created; you should see in the serial console at the bottom a message that reads "Server Started." Now that we know everything is good on the server side, let us go and test the client connection to the board. Open up your Web browser and navigate to the IP address of the Intel Galileo board followed by the port number which is ":3000." The URL should look similar to the following:

```
http://192.168.1.80:3000
```

You should now be connected to the server and a webpage will load with the words "Hello World" at the top of the page.

The Node.js framework has a built-in module called "HTTP" used in this code; remember we imported this in the first line of our code. We used this module to create a server on the Intel Galileo board. Then, in our code we checked for requests coming to the board from a client machine. Once we received the request, we simply then send a response with the words "Hello World."

Experiment 12: Creating a Web Server Using Express

Like we did in the previous experiment, we are going to create a Web server, but this time instead of using the HTTP module we are going to use the Express framework. Express is a Node.js Web application framework that provides more advanced features for creating more complex projects. We do not need to download or install anything at this point, as the Intel XDK will automatically download and install the Express module for us.

Open up a new project file in the Intel XDK and call it "ExpressWebServer." Type the following code into the main.js code:

```
var express = require('express');
var app = express();

app.get('/', function (req, res) {
  res.send('Hello World!');
});

var server = app.listen(3000, function () {

  var host = server.address().address;
  var port = server.address().port;

  console.log('Example app listening at http://%s:%s',
host, port);

});
```

Before we do anything, we must also modify the *package.json* file in our project. This file is automatically created by the Intel XDK and is used to see which modules need to be installed for the current project. The current package.json file looks like the following:

```
{
  "name": "blankapp",
  "description": "",
  "version": "0.0.0",
  "main": "main.js",
  "engines": {
    "node": ">=0.10.0"
  },
  "dependencies": {
  }
}
```

In the dependencies, add the Express module to look like the following:

```
{
  "name": "blankProject",
  "description": "",
  "version": "0.0.0",
```

```
    "main" : "main.js",
    "engines" : {
      "node" : ">=0.10.0"
    },
    "dependencies" : {
      "express" : "latest"
    }
}
```

As you can see, we define that we will use the most up-to-date version of the Express module. You can now upload the code to the Intel Galileo board by clicking the upload button in the toolbar. Now, you can click on the Build button on the bottom toolbar. Using the build button will pre-compile the code, but more importantly it will install the Express module that is required to run the project on the Intel Galileo board (see Figure 7.13).

When this is done, you can run the application again by clicking on the run button in the toolbar. You will then see the console spit out some information. After that you can open your Web browser and enter the Intel Galileo board's IP address, followed by the port number, which is set in the code :3000. You should see in your Web browser the words "Hello World."

As you can see, both Experiments 11 and 12 offer the exact same outcome but using much different modules. In this experiment, we

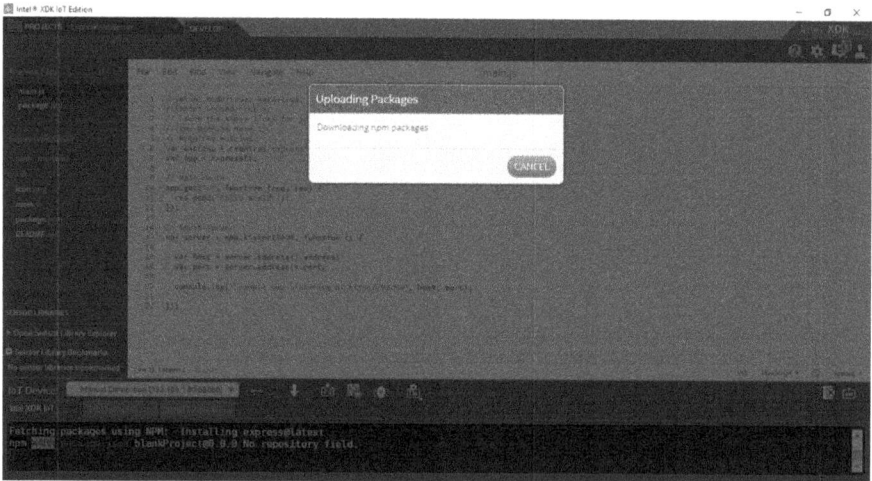

Figure 7.13 *Installing the Express module.*

used a more condensed code as opposed to the Node.js HTTP module in Experiment 11.

The Express module works by using routes. We therefore defined a path that we want to serve with our server, that is, we created a default route for our project with the following code:

```
app.get ('/', function (req, res) {
  res.send ('Hello World!');
});
```

You can see that the functionalities are the same as using HTTP module; however, using the Express module we can create more complex Web projects without using complicating coding. This somewhat simplifies the process and makes things easier to understand.

Experiment 13: Reading GPIO Pins on the Web Server

We are now going to put our Web server to good use reading sensor data from the GPIO pins on the Intel Galileo board, and then displaying this information on a webpage.

For this experiment, we are going to display the data from a light sensor, just like we did in Experiment 10.

Create a new project within the Intel XDK and load the following code into the main.js file:

```
var m = require ("mraa");
var util = require ('util');

var express = require ('express');
var app = express ();

var myAnalogPin = new m.Aio (1);
myAnalogPin.setBit (12);

app.get ('/read', function (req, res) {
  var myAnalogValue =  myAnalogPin.read ();
  res.send ("Analog pin 1 value is: " + myAnalogValue);
});
```

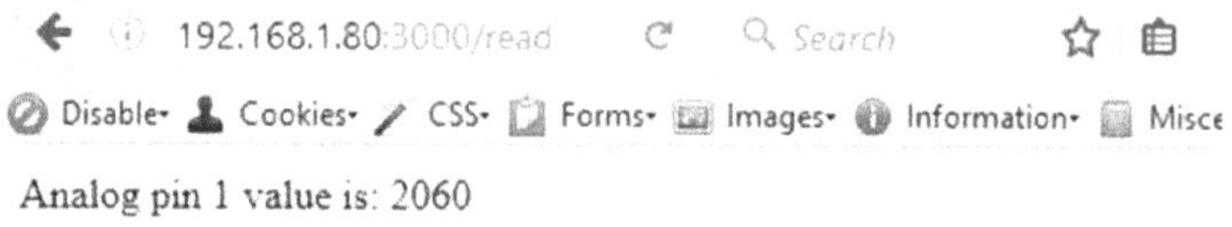

← ⓘ 192.168.1.80:3000/read ⟳ 🔍 Search ☆ 🗎

⊘ Disable• 👤 Cookies• ✎ CSS• 🗋 Forms• 🖼 Images• ⓘ Information• 🗔 Misce

Analog pin 1 value is: 2060

Figure 7.14 *Reading the analog light sensor through the Web.*

```
var server = app.listen(3000, function () {

  console.log("Express app started!");

});
```

Now that you have the code loaded, do not forget to add the Express module to the package.json file like we did in the previous experiment. Once this is done, upload the code and run the project using the Intel XDK. You should see a confirmation in the console that the project is running and the server is up. Then in your Web browser access the Galileo board by navigating to the following URL:

```
http://192.168.1.80:3000/read
```

Where the IP address is, insert the IP address of your own Galileo board; this will be different from the one above. Once the address is loaded, you should see the analog value of the light sensor that is connected to analog pin 1 on the Intel Galileo board, as shown in Figure 7.14.

In this experiment, we created a simple Web server using the Express framework, where we created a new route called "/read" that reads the state of the analog pin. When the client sends a request, the server sends the value to be displayed in your Web browser, using the following code:

```
app.get('/read', function (req, res) {
  var myAnalogValue =  myAnalogPin.read();
  res.send("Analog pin 1 value is: " + myAnalogValue);
});
```

Experiment 14: Controlling Digital Pins Using the Web

In the previous experiment, we learned how to read pins from the Intel Galileo board and display the values to the Web. This time we are going to control the pins on the Intel Galileo board to turn an LED on or off.

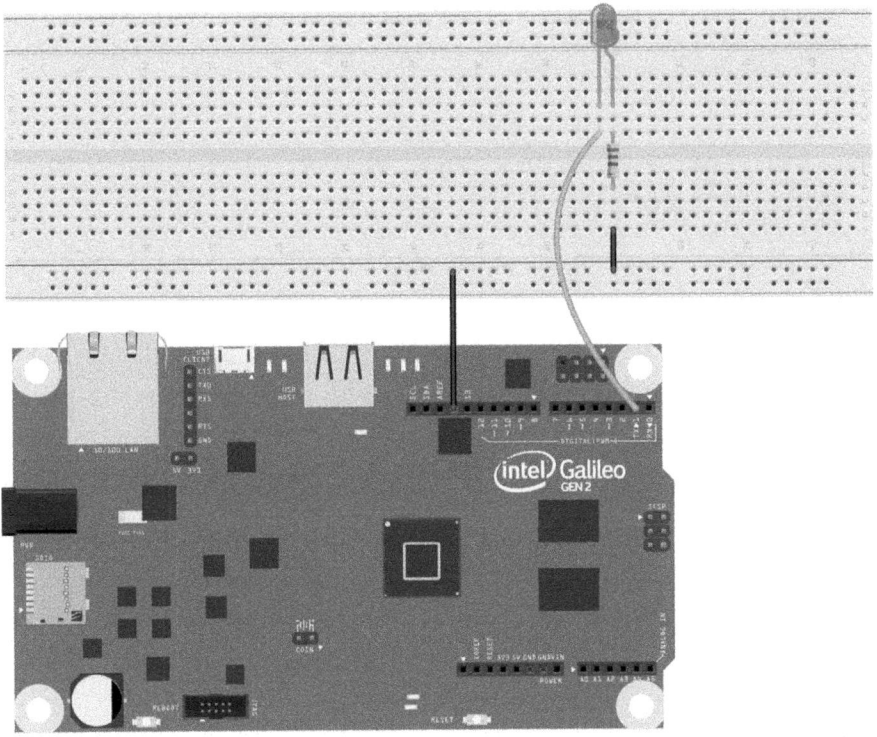

Figure 7.15   *Breadboard layout diagram for Experiment 14.*

Before we get started, connect an LED to one of the Intel Galileo's digital pins; for this experiment I will be using the Grove LED module and connect it to digital pin 1 using the Grove base shield. You can of course wire up your own LED using a breadboard by following the breadboard layout diagram in Figure 7.15.

In the Intel XDK, create a new blank project and load the following code into the main.js file:

```
var m = require("mraa");
var util = require('util');

var express = require('express');
var app = express();

var myDigitalPin = new m.Gpio(1);
myDigitalPin.dir(m.DIR_OUT);
```

```
app.get('/on', function (req, res) {
  myDigitalPin.write(1);
  res.send('Pin 1 is on');
});

app.get('/off', function (req, res) {
  myDigitalPin.write(0);
  res.send('Pin 1 is off');
});

var server = app.listen(3000, function () {
  console.log("Express app started!");

});
```

Now make sure that the Express module has been added to the package.json file, like in the previous two experiments. Once this is done, upload the project and run the application on the Intel Galileo board using the XDK. In the serial console, you should see confirmation that the server has started.

Open up your Web browser and navigate to the following URL to turn the LED on:

```
http://192.168.1.80:3000/on
```

You should get the message shown in Figure 7.16.

You should now see that the LED connected to the Intel Galileo board is on. To turn the LED off, simply do the same but this time use the "/off" route at the end of the URL:

```
http://192.168.1.80:3000/off
```

In this experiment, we used the "mraa" library to access the pins on the Intel Galileo board and the Express framework to create the Web server. We declared that digital pin number 1 is an output as shown in the following lines:

```
var myDigitalPin = new m.Gpio(1);
myDigitalPin.dir(m.DIR_OUT);
```

We then created a route on the server called "/on", which sets the state of digital pin 1 to HIGH or on:

```
app.get('/on', function (req, res) {
  myDigitalPin.write(1);
  res.send('Pin 1 is on');
});
```

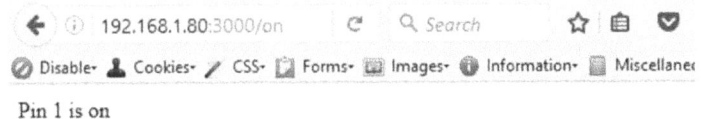

Pin 1 is on

Figure 7.16   *Turn the LED on in the Web browser.*

We did exactly the same to turn the LED off; we created a route called "/off":

```
app.get('/off', function (req, res) {
  myDigitalPin.write(0);
  res.send('Pin 1 is off');
});
```

You can create multiple instances of routes to determine what you want to happen. It essentially creates a function within the code when it catches the route of the URL.

Experiment 15: Home Automation

In this experiment, we are going to create a simple Web API to control a relay, which we will use to switch a watering pump on to water flowers. We will also create a user interface on the Web so that we can control the state of the relay with ease and get the value of a moisture sensor before we decide to water the flowers. In this experiment, we will also be introducing the *jade* module, which is a template engine for HTML.

For this experiment, we are going to use the Grove relay module and the Grove moisture sensor to detect the levels of moisture in the soil. The pump will be connected to the relay through a 9-V battery, which will drive the motor when it switches on. The hardware requirements can be found in Table 7.2.

Connect the Grove relay module to digital pin 1 on the base shield and also connect the moisture sensor to analog pin 1.

Now we can turn our attention to building the interface to control the relay switch. The on and off buttons will serve as the trigger controls from the Web to turn the relay on and off.

First, we need to create a structure for our project; we will use the Express module to do this. Using Express, we can separate the data from the interface. The measurement part to read the sensors and the state of the relay will stay in the Node.js main program. For the interface we will use the Jade file and then link up

Description	Appendix
Intel Galileo board	M1
Grove base shield	H3
Grove moisture sensors	R5
Grove relay	S1
6-V motor pump	H11
9-V battery	H5

Table 7.2 *Components and Hardware*

both the Node.js and the Express module to the Jade interface using some JavaScript.

In the main.js file, we need to include the Express module:

```
var express = require('express');
```

Then we create the application

```
var app = express();
```

We will now set Jade as the view engine:

```
app.set('view engine', 'jade');
app.set('views', __dirname + '/views');
```

Here, __dirname is the root directory of the app. In the next line, we define where the JavaScript files will be stored:

```
app.use(express.static(__dirname + '/public'));
```

Now, we will define the main route where the interface can be accessed:

```
app.get('/', function(req,res){
  res.render('interface');
});
```

The API we are creating will be accessed later by the interface, using JavaScript.

For example, for reading the moisture sensor, we define the API access as follows:

```
app.get('/api/moisture', function(req,res){

var b = moist_sensor_pin.read();
console.log("Analog Pin (A1) Output: " + b);
var moisture = (((4096-b)/4096)*100);
console.log(moisture);
```

```
  json_answer = {};
  json_answer.moisture = moisture;
  res.json(json_answer);
});
```

We then start the app at the following point:

```
var port = 3000;
app.listen(port);
console.log('Listening on port ' + port);
```

We modify the package.json file at this point as we need to include all of the Node.js modules that are used by our app:

```
"dependencies": {
      "util": "latest",
      "express": "latest",
      "jade": "latest"
   }
```

Now let us turn our attention to the interface.jade file. In the toolbar go to File > New; this will create a new file that you can save as interface.jade in a new folder called "Views." We then define the containers that will hold the different elements of the interface:

```
html
   head
     title Watering Flowers Interface
     script(src='https://code.jquery.com/jquery-2.1.1.
min.js')
     script(src='/js/interface.js')
     link(rel='stylesheet', href="https://maxcdn.boot-
strapcdn.com/bootstrap/3.3.0/css/bootstrap.min.css")
   body
     .container
       h1 Water the flowers
       h3.row
         .col-md-4
           div
         .col-md-4
           div#moisture Moisture Level:
       h3.row
         .col-md-4
           div Relay control
         .col-md-4
           button.btn.btn-lg.btn-block.btn-primary#on On
```

```
.col-md-4
    button.btn.btn-lg.btn-block.btn-danger#off Off
```

Now we can create a new JavaScript file, which will refresh the container's variables by calling the API in our main.js file:

```
$( document ).ready(function() {

    $.get('/api/moisture', function(json_data) {
        $('#moisture').html('Moisture Level: ' + (json_
data.moisture).toFixed(2) + ' %');
    });

    $('#on').click(function() {
        $.get('/api/relay?state=1');
    });

    $('#off').click(function() {
        $.get('/api/relay?state=0');
    });

});
```

Now we can upload the code to the Intel Galileo board; it will download all the packages required as defined in the package.json file. Click run and you should see in the console that the server has started. Now open up your Web browser and navigate to the following URL:

```
http://192.168.1.80:3000
```

Where the IP address is, insert the IP address of your Intel Galileo board. You should see something similar to Figure 7.17.

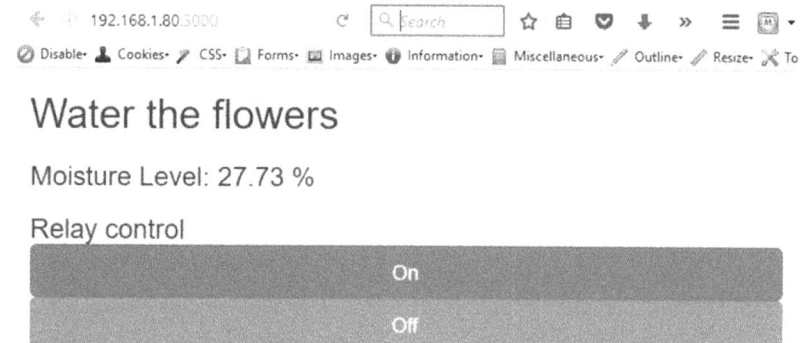

Figure 7.17 *Web interface.*

You should see on the webpage the current moisture sensor level as well as two buttons for turning the pump on and off.

Summary

In this chapter, we have learned how to send our data from connected hardware to the Web using dweet.io. We have also learned how to create a local Web server using the Express module and control our hardware devices using the Web. You should now have the skill and ability to go on to create your own IoT projects using the Intel Galileo board. In the next chapter, we will look at some tools and tips to help you create your own projects.

8

Tools and Tips

Even though this book primarily focused on the use of the SeeedStudio Grove modules, you may wish to add your own components to the Intel Galileo board instead. This section gives you some useful tips for creating your own projects and on how to best utilize your resources. Starting your own projects can be a bit daunting at first and sometimes can seem frustrating and complicated; this useful information should help you on your way.

Breadboards and Prototyping Boards

A breadboard (see Figure 8.1) is usually a rectangular plastic ABS box with lots of little holes in it; the holes are contacts in which you can insert electrical components or wires with ease. Breadboards are often used to put together a concept design of a circuit without having to solder any components. Instead, you just poke the wires or legs of the component into the holes, creating a contact. The contacts are usually arranged in rows by connecting the metal contacts underneath the breadboard. The best thing about using a breadboard is that you can change the circuit design at any point, so that you can replace or rearrange components with ease without having to solder/desolder any joints.

When you place components in a breadboard, not much happens unless you connect jumper wires to create an electrical circuit (see Figure 8.2). Wire used in electronics is copper surrounded by outer plastic insulation, and is usually called hook-up wire. Wire comes in all sorts of diameters, often referred to as gauge; the standard measurement in the United States is AWG. It is always advisable to use solid wire rather than stranded wire because solid wire inserts into the breadboard much easier than does stranded wire. If you are lucky, your electronics shop

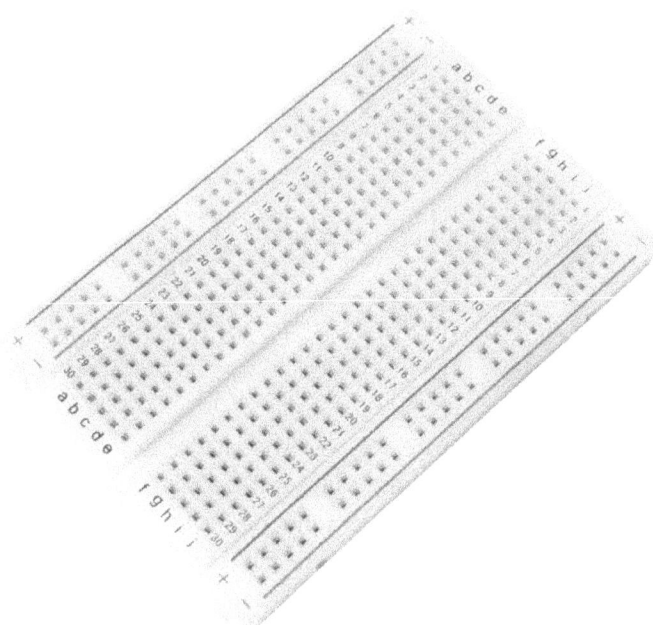

Figure 8.1 *Breadboard.*

Figure 8.2 *Jumper wires.*

will sell jumper wires, which are short lengths of wire with a single pin on each end.

If you create your circuit design on a breadboard, you may decide that you want to make it permanent by soldering components in place on a printed circuit

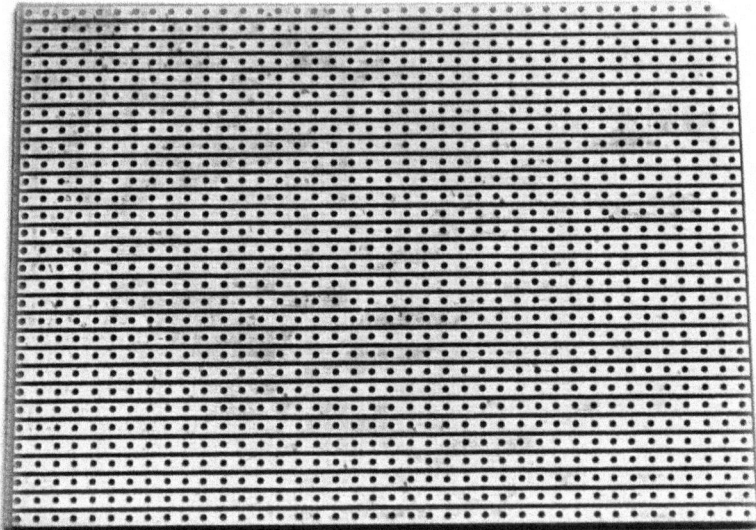

Figure 8.3 *Prototyping board.*

board (PCB). To do this, you may have to get a universal PCB, which in some ways is similar to a breadboard layout. A prototyping PCB (see Figure 8.3) has rows of individual holes across it, much like all the pins on a breadboard. Generally, all the components will go on the top of the board, and you solder underneath; when you solder all the wires, they usually go underneath also. This makes a much neater and cleaner board to work with and can help avoid congestion if you are soldering a lot of components.

Multimeter

A multimeter is a useful device that measures electricity, just like you would use a ruler to measure distance or a stopwatch to measure time. The best thing about a multimeter is that it measures many different things, such as voltage, current, resistance, and much more. A standard multimeter will have a large dial in the middle, which lets you select what you want to measure (see Figure 8.4).

Most multimeters can measure voltage, current, and resistance; some multi-meters also have a continuity check, which tests to see if the electrical circuit is complete by producing a loud beep when two things are electrically connected. This is helpful in diagnosing problems with a circuit—you can trace the voltage

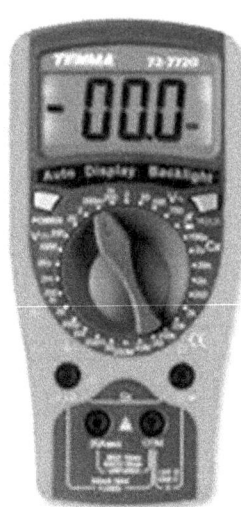

Figure 8.4 *A typical multimeter.*

around the circuit and find which part is incomplete or not functioning the way it should. In contrast, you can make sure that two things are not connected just in case you do not want a certain part of your circuit to short, or you may want to test your soldering skills by not accidentally soldering joints together.

There are some advanced multimeters, which are often expensive, that have certain additional functions, such as the ability to measure transistors or capacitors. These functions are more suited for the professional engineers who design and manufacture high-end products.

In order for us show to you how a multimeter works, it is important to understand what we are measuring:

- Voltage is how hard the electricity is being pushed through a circuit—a higher voltage is being pushed through a circuit really hard. Voltage uses the symbol V.

- Current is how much electricity is flowing through a circuit—a high current indicates that more electricity is flowing through. The symbol for current is A.

- Resistance is how difficult it is for electricity to flow through a circuit—a higher resistance indicates that it is much more difficult for it to flow through. Resistance is measured in ohms, and the symbol for ohms is Ω.

It is worth noting that the symbol used for a unit in general may differ from the symbol used for that unit in a variable equation.

Soldering

Learning how to solder is an essential skill in the world of electronics. Although you can get by just using a prototyping board, you may still need to solder headers onto the board or make some small modifications to a component.

Solder refers to the alloy that typically comes on a wire spool or tube (see Figure 8.5), and it is this solder that we use to fuse components together on a printed circuit board. Solder usually comes in two types: leaded and lead-free. When solder was first around, it was generally made up of an alloy using both lead and tin; since then, it has become known that lead can be quite harmful when exposed to in large amounts. Lead was used in solder because it has a great low melting point and it created really good solder joints, which produced a highly reliable circuit board. Unfortunately in the European Union, leaded solder is not Restriction of Hazardous Substances (RoHS) compliant, and this restricts the use of leaded solder in electrical equipment, hence is why lead-free solder is commonly used. Lead-free solder is usually made up of metals such as silver and copper. Lead-free solder does come with downfalls—for example, it has a much higher melting point because of the tin content and as such requires a high-powered soldering iron.

Figure 8.5 *Solder spool.*

Lead-free solder often contains a flux core, which helps give the same quality effect as leaded solder. Flux is a chemical agent that aids in flow and creates much better contacts when finished.

Many tools are used to aid in soldering, but none are more important than a soldering iron. Soldering irons come in a variety of types, ranging from basic soldering irons to complex soldering stations, but they all serve the same function and purpose. Usually, a good place to start is to buy a station that comprises a soldering iron, either a digital or analog controller, and a stand. These stations are becoming more common now and can be inexpensive to purchase at your local store.

Over time your soldering tip will start to oxidize and will turn black. This is bad because the soldering iron will not cling to the solder and you will not be able to solder a component. This is more commonly found with lead-free solder. This is where a simple soft sponge comes to the rescue—every so often you should clean the tip by wiping all the excess oxidation off. For even better results, you can use a brass wire sponge (Figure 8.6).

Apart from a soldering iron and solder, several other great accessories can aid in the process of soldering. A solder wick is a vital tool for mopping up if you have made a bit of a mess; you can also use it for desoldering. Solder wick is made up of thin copper braiding, and just like any PCB, it will soak up the solder, erasing any excess drops.

Figure 8.6 *Brass sponge.*

Figure 8.7 *Tip tinner.*

You can also use a tip tinner to clean your tip (see Figure 8.7). It is composed of a mild acid that helps remove any unwanted residue left on your soldering tip and prevents oxidization when the tip is not in use.

As previously mentioned, some lead-free solder comes with a core flux; however, sometimes it is not enough and extra flux may be required. Flux pens are used to allow difficult components to create a better bond to the PCB.

Analog versus Digital

Analog and digital signals are used to transmit an array of information, usually conveyed through electrical signals. The main difference between the two signals is that analog signals are transmitted in pulses of varying amplitude, and digital signals are transmitted into a binary format such as a one or zero, where each bit represents a distinct amplitude.

Analog refers to the circuits in which quantities such as voltage or current vary at a continuous rate over a period of time. Electronic signals represent information by changing their voltage or current over time. The signal takes any value in a given range, and each signal value can represent a different kind of information. Any change that takes place in the signal has a significant impact on the overall result.

Something important to take into account is that analog signals can create noise, which is classified as a disturbance or variation, which can be caused by thermal vibrations. Because any slight variation in the signal can affect the outcome, this noise can have a significant effect, especially over long distances as the signal degrades.

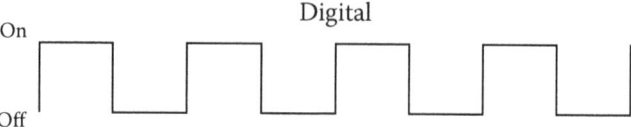

Figure 8.8 *Digital signal.*

When designing a system, analog circuits are much harder and complex and require more skill compared with digital systems. This is primarily why digital systems have become more common; in addition, they are much cheaper to manufacture.

Digital systems are much easier to understand—they do not use a continuous range like analog, so any noise or slight variation in the signal does not affect the result of a digital signal. Digital systems generally have only two states, and they are represented by two different voltages—usually 0 is equal to ground and 1 is equal to +V. (See Figure 8.8.)

The main advantage of using a digital system is that when compared with analog, the signal does not degrade over time and it can be quite easily replicated without any loss. The main disadvantage is that digital circuits consume much more power than do analog circuits, and when circuits consume more power, that generally means more heat, which in turn increases the complexity of designing a circuit.

Suppliers

All of the parts used in this book are easily available for purchase from a number of stores through the Internet. However, sometimes it can be quite difficult to find exactly what you are looking for, especially in your own country. Finding suppliers in your own country can reduce the cost of shipping, offering you a more affordable solution.

Most of the parts used in this book can be purchased through SeeedStudio or one of their global distributors such as Digi-Key Electronics. Most of the components are also available from many other suppliers, such as Adafruit Industries and SparkFun. These companies often specialize in the maker market and manufacture their own products.

Digi-Key Electronics, one of the largest component distributors in the world, is based in North America. Digi-Key stocks a wide range of components for manufacture while also stocking maker products from the likes of Adafruit, SparkFun, and SeeedStudio. Digi-Key offers great distribution across the globe, making it a one-stop shop for all your maker needs. Digi-Key also offers a number of tutorials and guides through maker.io where you can find hobbyists.

SparkFun, based in the United States, is an online retail store that sells various pieces to make your electronic projects. In addition to their products, they provide classes and online tutorials designed to help educate individuals in embedded electronics (www.sparkfun.com).

Adafruit Industries was founded in 2005 by MIT engineer Limor Fried. Adafruit was designed to create the best place online for learning about electronics and making the best-designed products on the market. Adafruit designs and develops their own products in house; they sell these on their website, along with several tutorials to get you started (www.Adafruit.com).

SeeedStudio is based in Shenzhen, China, and the United States. Having access to the Far East market makes them a prime supplier for all hardware components. Not only does Seeed have a wide range of electrical components at very low cost, they also offer a wide range of services, from manufacturing to 3D printing to laser cutting and more. Check out their website for further information (www.seeedstudio.com).

A

Components and Hardware

The hardware and component table for each project lists appendix codes for each component that is used. This appendix lists all the parts and offers some sources as to where they can be purchased. (A couple of Galileo kits and modules are listed in Table A.1.)

Resistors

Resistors are low-cost components—almost less than one cent each. Often suppliers will sell them in packs of 50 or 100. There are common resistors that get used a lot, such as 220R, 270R, 1K, and 10K values, so it can be useful to keep several of these values on hand (Table A.2).

After a while you might find yourself buying a lot of resistors, and in some cases it is better to buy them in kits, which stock the most popular values used in everyday electronics.

Code	Description	Source
M1	Intel Galileo Gen 2 board	Digi-Key: 1050-1051-ND
M2	Grove Starter Kit Plus	SeeedStudio: 110020002

Table A.1 *Galileo Kits and Modules*

Code	Description	Source
R1	Grove potentiometer	Included in Grove kit Digi-Key: 1597-1093-ND
R2	Grove photocell	Included in Grove kit Digi-Key: 1597-1125-ND
R3	Grove NTC temperature sensor	Included in Grove kit Digi-Key: 1597-1134-ND
R4	Grove air quality sensor	Digi-Key: 1597-1162-ND
R5	Grove moisture sensor	Digi-Key: 1597-1141-ND

Table A.2 *Resistors and Sensors*

Some companies that sell resistor kits are listed below:

+ SeeedStudio: 110990043
+ SparkFun: COM-10969
+ Digi-Key: 110990043-ND

Switches

Switches are great for turning things on or off such as LEDs and lights. Switches come in many different form factors and in varying sizes. For most projects you should be familiar with using tactile switches, but essentially they all work the same and have the same outcome (Table A.3).

Diodes

LEDs are the bread and butter of visual feedback. They come in all sorts of different shapes, sizes, colors, and voltages and are one of the most common components you will use when you start out tinkering. Most likely you will use these in almost every project (Table A.4).

Code	Description	Sources
S1	Grove relay	Included in Grove kit Digi-Key: 1597-1327-ND
S2	Grove tactile button	Included in Grove kit Digi-Key: 1597-1285-ND

Table A.3 *Switches*

Code	Description	Sources
D1	Grove LED	Included in Grove kit Digi-Key: 1597-1338-ND (RED) Digi-Key: 1597-1340-ND (GREEN) Digi-Key: 1597-1342-ND (BLUE)
D2	Low-voltage lamp	Most maker stores
D3	Grove RGB LED	Digi-Key: 1597-1339-ND

Table A.4 *Diodes*

Hardware and Miscellaneous

Most hardware and in particular some miscellaneous parts can be found in most maker/hobbyist stores worldwide (Table A.5).

Code	Description	Sources
H1	Jumper wires	Digi-Key: 377-2093-ND
H2	Digital multimeter	Digi-Key: PRO-50A-ND
H3	Grove base shield	Included in Grove kit Digi-Key: 1597-1189-ND
H4	Grove LCD module	Included in Grove kit Digi-Key: 1597-1336-ND
H5	9-V PP3 battery	Most electrical stores
H6	Servo motor	Included in Grove kit Digi-Key: 1528-1076-ND
H7	Intel Wi-Fi mini-PCIe module	Amazon/eBay
H8	Dual-band antennas	Digi-Key: WM5030-ND
H9	Full-size PCIe bracket	Amazon/eBay
H10	4-GB (min) micro-SD card	Included in Grove kit (8 GB) Digi-Key: AF4GUD3A-OEM-ND
H11	9-V DC motor pump with tubing	Digi-Key: 114990073-ND

Table A.5 *Hardware and Miscellaneous*

INDEX

Note: Page numbers in *italics* indicate figures and tables.

Lightning Sou
Milton Kev
UKHW020
351443U   3/P

9 781259 644795